GIS Basics

Stephen Wise

London and New York

First published 2002
by Taylor & Francis
11 New Fetter Lane, London EC4P 4EE

Simultaneously published in the USA and Canada
by Taylor & Francis Inc,
29 West 35th Street, New York, NY 10001

Taylor & Francis is an imprint of the Taylor & Francis Group

© 2002 Stephen Wise

Typeset in Sabon by
Integra Software Services Pvt. Ltd, Pondicherry, India
Printed and bound in Great Britain by
The Cromwell Press, Trowbridge, Wiltshire

British Library Cataloguing in Publication Data
A catalogue record for this book is available
from the British Library

Library of Congress Cataloging in Publication Data
A catalog record has been requested

ISBN 0–415–24651–2 (pbk)
ISBN 0–415–24650–4 (hbk)

Contents

Illustrations

Preface

Geographical Information Systems are now widely used in areas such as commerce, local government, defence and academic research. There are numerous introductory textbooks and many Universities provide introductory courses for undergraduates, often as part of a degree in subjects such as Geography, Planning or Archaeology. There is thus a wealth of material available for learning about GIS – how to use one to solve spatial problems, which organisations are using them and for what applications, the issues relating to access to data and the use of personal information. Anyone taking a course or reading one of the textbooks will be familiar with ideas such as the use of buffers to define zones of influence, or operations such as polygon overlay. But have you ever wondered just how the software can perform these operations? Just how does a GIS work out which points fall inside a buffer for example?

Those who want to extend their understanding by finding out how a GIS actually works will often find that it is difficult to find any suitable instructional material. Many of the software developments which have lead to the creation of GIS software are made by computer scientists, and although texts do exist which explain these developments, they are not easy to understand unless you have some grounding in mathematics and computer science.

This was the problem I faced when trying to design an advanced GIS course for third year undergraduates. I felt it was important that they had some understanding of the fundamental concepts underlying GIS software – how lines can be stored in a vector GIS for instance, and how to determine whether two lines intersect – but there was very little I could recommend for them to read and certainly nothing in the way of a single coherent introduction. This was my motivation for writing a series of articles in GIS Europe, under the title 'Back to Basics'. There were 20 in all, and they attempted to explain some of the fundamental ideas of computer science in simple language to those with no previous experience of this subject, and with little mathematical background. This book is based on these articles, but extended to provide a more comprehensive coverage of the area. The aim is to introduce some of the main ideas and issues in the design of GIS systems, not to provide comprehensive coverage of all the data structures and algorithms used in current systems.

It is assumed that the reader already has some knowledge of Geographic Information Systems. This book will not provide a general introduction to GIS, but will focus solely on technical issues relating to the inner workings of GIS. No previous experience of computer science is assumed – indeed, it is hoped that the book will serve as a gentle introduction to some of the main issues in Computer Science, and help make some of the GIS literature which comes out of that discipline more accessible. The book contains some mathematics, but this is not at a very advanced level, and all the ideas can be understood without following the details of the mathematics.

Steve Wise
Bakewell
October 2001

Acknowledgements

This book has its origins in a series of articles in GIS Europe which appeared under the title 'Back to Basics' so thanks must go to Vanessa Lawrence, for agreeing to publish them, and staff at GIS Europe for their help in this process. Thanks to Tony Moore at Taylor and Francis who first suggested I turn the articles into a book and then enlisted the able assistance of Sarah Kramer to make sure I got it done! The quality of the book was improved enormously by extremely helpful suggestions from Peter van Oosterom and Nick Tate who both commented on an early draft.

The maps of Chatsworth maze in Chapter 1 are reproduced with the kind permission of the Chatsworth House Trust. Many of the figures in Chapters 9 and 10 are derived from Ordnance Survey digitized contour data and are reproduced with kind permission of Ordnance Survey © crown copyright NC/01/576.

Finally, a big thanks to the forbearance of my family – Dawn, Gemma and Hannah – who will see more of me now that this is done. Whether or not that is a good thing only they can say!

1 Introduction

Computer software is generally designed to undertake particular tasks or to solve problems. Geographic Information Systems (GIS) are no exception, and of course in this case the tasks or problems involve spatial data. One of the most intriguing things about spatial data problems is that things which appear to be trivially easy to a human being can be surprisingly difficult on a computer. This is partly due to the complex nature of spatial data, and partly to the way that computers tackle problems.

If you are reading this book, you should be familiar with the way that GIS can be used to answer questions or solve problems involving spatial data. You will also be aware that in order to use a GIS you have to learn how to apply the tools available in a GIS to your particular needs. This often involves rethinking the problem into a form which can be tackled using a GIS. For example, you may be interested in whether traffic has an effect on the vegetation near roads. This is a spatial problem, but to use GIS to study it we will have to be more specific about what sort of effect we might expect, and what we mean by 'near'. We might decide to concentrate on species diversity using results from a series of sample plots at varying distances from a road. A very simple analysis would be to compare the results for plots within 10 m of the road with those which are furtheraway than this. In GIS terms, we would create a 10-m buffer around the road and produce summary statistics for sites within and for those outside it.

In the same way that learning to use a GIS involves learning to 'think' in GIS terms, learning about how GIS programs work, involves learning to 'think' like a computer. Given a map showing the buffer zone and plots from the previous example, most people would have little trouble in identifying which points fell inside the buffer. However, it is not a trivial task to program a computer to do the same. The purpose of this book is to provide an introduction to what is involved in producing software which can solve spatial problems. In order to do this we will need to learn how computers 'think' – how they solve problems.

1.1 HOW COMPUTERS SOLVE PROBLEMS

The problem we will consider is that of finding a route through a maze, such as the one shown in Figure 1.1 (which is actually the maze at Chatsworth House in Derbyshire). This is quite a simple maze, and most people will be able to trace a route to the centre fairly easily.

Part of the reason that this is a simple task, is that we can see the whole plan of the maze laid out before us. Now imagine that you are standing in the gardens of Chatsworth House, at the entrance to the maze. All you can see is the opening in the hedge. When you enter you will be at a junction with a path leading to the left and a path to the right. It is clear that if you are going to find your way to the centre, you will have to adopt some form of strategy. With many mazes, a method which works is to place your hand on one of the walls, and walk along keeping your hand on the same wall all the time. If you place your left hand on the left hand wall in the Chatsworth maze then the route you take will be as shown in Figure 1.2.

This strategy has worked – we have found a route to the centre. What is more, in order to make it work we only have to process one piece of information at a time – does the wall on the left carry straight on or turn a corner? This is important because computers operate in this one-step at a time fashion.

This is not the place for a detailed description of how computers work but it is important to understand this feature of them. Large amounts of information can be held, either on disk, or in the computer's memory, awaiting processing. The actual processing is carried out by what is called the Central Processing Unit or CPU for short which can perform a number of operations on the information which is passed to it. Perhaps, the most

Figure 1.1 The maze at Chatsworth House.

Figure 1.2 Route through the maze using the left hand rule.

basic is addition – the computer passes two numbers to the CPU which adds them and returns the answer to be stored in memory or on a disk file (this is an oversimplification but will suffice for our purposes). Perhaps, the most surprising thing is that the CPU can only deal with two pieces of information at a time. If we want to add together three numbers, we have to add the first two, save the result and then add this to the third one. So how can a computer perform complex calculations? By breaking them down into a sequence of simpler calculations. And how can it solve complex problems? By breaking them down into a sequence of simpler steps. The resulting sequence of steps is called an algorithm.

Our maze solving strategy is a good example of an algorithm. It is designed to solve a problem, and does so with a series of steps. What is more, the steps are unambiguous – there is no subjective judgement to be used, simply a set of rules to be blindly followed. The maze example is also useful to illustrate some other important ideas in algorithm design.

The first is that this simple algorithm will not always work. It is very easy to design a maze which cannot be solved by this algorithm, and an example is shown in Figure 1.3.

Here our simple algorithm will simply circle around the outer wall and return to the entrance. We will need a more sophisticated strategy. For instance, we will need to be able to detect that we have returned to the same place, to avoid circling around the outer wall for ever. This raises another issue – in order to solve a problem on the computer, we need some way of storing the relevant information in the first place. There are two stages to this. The first is to create a representation of the information which will tell us what we want to know and which we can store on a computer. In the case of the maze, the obvious thing would be to store the plan of the

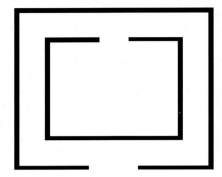

Figure 1.3 A maze which cannot be solved using the simple algorithm.

walls, as a set of lines, and this could certainly be done. However, what is actually important is the paths and particularly where they join one another – a diagram of the paths for Figure 1.3 might look like this.

Both Figure 1.3 and Figure 1.4 are examples of what are sometimes loosely referred to as 'Data Models' (a more thorough description of data modelling is presented in the next section). These are still pictorial representations though, and so the next stage would be to work out a method for storing them in a computer file, which requires what is sometimes called (again rather loosely) a data structure. It is the data structure which the computer will use in order to solve our problem, using whatever algorithm we eventually come up with. For instance, a better algorithm than our wall-following approach would be to use the path data model. Using this we will need an algorithm which traces a route from the origin point to the destination. At each junction, we will need to decide which path to try first. We will also need to keep track of which paths have been tried, to avoid trying

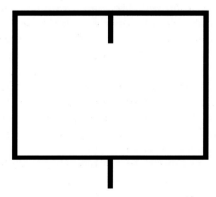

Figure 1.4 Diagrammatic plan of the paths in Figure 1.3.

them again, so the data structure will need to be able to store an extra piece of information for each path, labelled 'Has this been tried or not'? It can be seen from this that the design of data structures and algorithms are closely linked.

The key developments which have lead to the creation of modern GIS software were the design of data structures and algorithms for handling spatial data. The maze example has shown that algorithms are sequences of steps designed to solve a problem, where the information for the problem is stored in a suitable data structure. There is one other important feature of algorithm design, which is nicely illustrated by the original solution to the Chatsworth House maze (Figure 1.1). This algorithm worked but it was very inefficient since it involved walking up several paths, in most cases twice. With a small maze this would not be too important, but with a much larger one, it would clearly start to matter. From this we can see that algorithms should not simply work, but work efficiently.

This first section has provided an informal introduction to the main themes which will be covered in the book – the design of data structures and algorithms for handling spatial data. Before these can be covered however, we need to consider the important topic of data modelling in a little more detail.

1.2 HOW COMPUTERS STORE SPATIAL DATA: VECTOR AND RASTER DATA MODELS

We have seen that in order to process information about the real world on the computer, we must have a representation of that information in a form which the computer can use. The real world is infinitely complex, so we will first need to select the information which is relevant to the problem in hand and then decide how to organize this, to be stored on the computer. This process of going from some real world phenomenon to a computer representation of it is termed 'data modelling' and it can be considered as a series of steps as shown in Figure 1.5 which is based on Worboys (1995).

We have already seen examples of the first two stages in this process with the example of the maze problem. The original map of the maze was an example of an Application Domain Model – a representation of the system of interest in a form comprehensible to people who are familiar with the real world system. It is a simplified model of the real maze and like all models it represents the most important elements, in this case the walls and paths, leaving out unimportant details, like the height of the hedges or the type of surface on the paths.

For the maze solving problem, this model could be simplified even further since all that was needed was the connections between the paths and these could be represented as shown in Figure 1.4. This is an example of what mathematicians call a graph, and is an important way of representing

Model	Characteristics	Example
Application Domain Model	Not a computer model but capable of conversion to one	Map

⇩

| Conceptual Computer Model | Hardware independent, Software independent | Vector model of map |

⇩

| Logical Computer Model | Hardware independent, Software dependent | Digitised map in ARC/INFO ASCII format |

⇩

| Physical Computer Model | Hardware dependent, Software dependent | ARC/INFO coverages on Unix workstation |

Figure 1.5 Data modelling steps after Worboys (1995).

spatial data as we shall see in Chapter 2. In effect, this is a model of one element of the original map, and forms the Conceptual Computational Model in this example. It is a more abstract representation than the original map, but it is not in a form which can be directly stored in a computer.

The next two stages in the process will be described in a moment, but first a word about these first two stages. In order to be useful for handling a wide range of spatial data, GIS must be able to represent a wide range of Application Domain Models and Conceptual Computational Models. In practice most current GIS systems were essentially written to deal with just one Application Domain Model – the paper map. This is a slight over-simplification, but it is certainly true that like our maze map, maps of all kinds are models of the real world which take certain elements of the natural or built environment and provide a simplified representation of them. Many of the key developments in GIS have involved the design of the conceptual computational models and logical computational models for representing spatial data derived from maps on the computer. Many current developments in GIS are concerned with extending GIS software to be able to deal with those things which are difficult to represent on a map – change over time, or 3D variation for example, and current data modelling methodologies certainly do not restrict themselves to maps. However, for the purpose of the first part of this book it is a useful simplification to focus solely on the data structures and algorithms used for handling map-based data. The issue of other types of spatial data will be addressed in due course, and the further reading section includes references to works which give a fuller account of data modelling. So to return to our description of data modelling, the Application Domain Model for GIS can be considered to be the map, of which there are two main types. The topographic map is a general purpose map showing a range of features of the Earth's surface – examples are the Ordnance Survey Landranger and Pathfinder maps in the United Kingdom, or the sort of maps produced for tourists showing places of interest, roads, towns etc. The other type of map is the thematic map, of which a soil map is a good example. This is not a general map of the area but a map of the variation or pattern of one thing or 'theme'. Another good example is maps derived from census information showing how the characteristics of the population vary across an area.

In both cases, two types of information are being shown on the map. First, the map shows us where things are – the location of settlements, woodlands, areas of high population density for example. Second, it shows us some of the characteristics of those things – the name of the settlement, whether the woodland is deciduous or coniferous, the actual number of people per hectare. The location is shown by the position of the features on the map, which usually has some kind of grid superimposed so that we can relate these positions on the map to real world locations (for example measured in latitude and longitude). The characteristics of the features are usually indicated by the map symbolism, a term which includes the colours used for lines or areas, the symbols used for points and filling areas and whether the line is dashed or dotted. Hence, the settlement name would be printed as text, the woodland type would usually be indicated by using a standard set of symbols for each woodland type, and the population density would be shown by a type of shading with the actual values indicated on a key.

In order to store this spatial data, the computer must be able to hold both types of information – where things are and what they are like, or to use the correct terms the locational and attribute data. This requires an appropriate Conceptual Computational Model, and in fact two are in common use – the Vector and Raster models. Much of what follows will be familiar to most

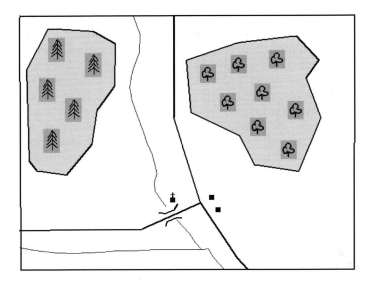

Figure 1.6 Imaginary topographic map.

readers, but it forms a necessary prelude to the description of the Logical Computational Models which follows. To illustrate the principles of the vector and raster models, let us take the example of the topographic map shown in Figure 1.6, and see how the information shown on it would be stored in each case.

In a raster GIS, an imaginary grid is laid over the map. Each cell in the grid, called a pixel, is examined to see what feature falls within it. Since computers work best with numbers a code is used to represent the different features on the map – 1 for coniferous woodland, 2 for deciduous woodland etc. Therefore, all the pixels which cover areas of coniferous woodland will have a number 1 stored in them. Figure 1.7 shows this process for part of the map – on left a grid is superimposed on the map, on the right is the grid of numbers. Blank cells in the grid would actually have a zero stored in them, but these are not shown here.

The end result of this will be a grid of numbers which can easily be stored in a computer file. The grid can be used to redraw the map – each pixel is coloured according to the code value stored in it – dark green for conifers, light green for deciduous woodland and so on – producing a computer version of the original map. We can also do simple calculations using the grid – by counting the number of pixels containing deciduous woodland we can estimate how much of the area is covered by this type of landcover.

In assigning the codes to the pixels, there will be cases when one pixel covers more than one feature on the original map. This will occur along the

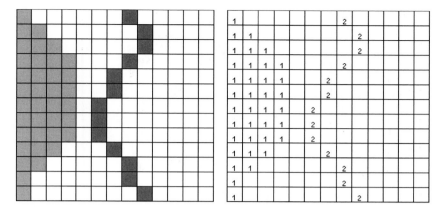

Figure 1.7 Raster version of part of map in Figure 1.6. Left – diagrammatic view of raster map. Right – the grid of numbers which actually represents the map in a raster system.

boundaries between landuse classes for example, in these cases, whichever item occupies the majority of the pixel will have its value stored there. Another example is when a road and a river cross – we cannot record both in the same pixel. In this case, rather than try and store all the data in a single grid, we use separate grids for different types of feature, so that in this case roads and rivers would each be stored in their own grid. Note that this means that we can use the GIS to look for places where roads and rivers cross, by laying one grid on top of the other. The end result of this process is a grid or a series of grids of numbers which represent the features on the map – one grid is often referred to as a layer. The numbers in the pixels therefore represent the attribute data but how is the location of the features stored in this kind of system? Since the pixels are all square, as long as you know the location of one point in the grid, and the size of the pixels, the location of any other pixel can be worked out fairly simply. Most raster systems therefore store this information for each raster layer, sometimes in a special file which is kept separately from the layer itself.

In a vector GIS, each object which is to be stored is classified as either a point, a line or an area and given a unique identifier. The locational data from the map is then stored by recording the positions of each object, as shown in Figure 1.8 which shows the vector locational data for the map in Figure 1.6.

The buildings have been stored as points and the location of each one is recorded as a single pair of coordinate values. It is clear that we have only stored part of the information from the map – for instance two of these points are houses, while the third is a church. This attribute information is normally stored separately, in a database, and linked to the locational data

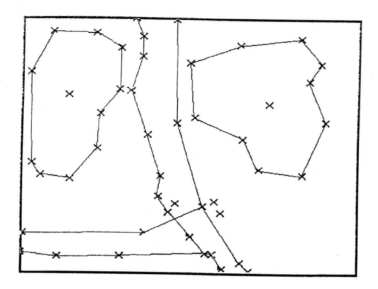

Figure 1.8 Vector version of topographic map in Figure 1.6.

using the identifier for each point. Lines, such as the roads and rivers, need to have multiple coordinates representing points along the line, except in special cases where it may be possible to use a mathematical function to represent the course of the line. Areas, such as the two woodlands, are represented using a line to represent the boundary, and a point marking some point inside the area. This is not essential, but is useful for locating labels when the area is drawn.

Vector and raster, which are often loosely referred to as Data Models, could more properly be regarded as Conceptual Computational Models using the terminology of Figure 1.5. This means each provides a model of a map in a form which can be stored on the computer, but without being tied to any particular software package or type of computer. This is the purpose of the next stage in the modelling process – the design of a Logical Computational Model. As a simple example of what is meant by this term, consider the example of a raster layer, which consists of a set of values in regular grid representing values of some phenomenon across space. Almost all computer languages have a data structure called an array, which can be used to hold 2D matrices of values – for example in FORTRAN, an array could be set up and filled with values read from a file as follows:

```
INTEGER LANDUSE(100,100)
OPEN (UNIT=1,FILE='LANDUSE.DAT')
READ(1,*) LANDUSE
```

The FORTRAN array is therefore one example of a Logical Computational Model – it is a representation of the raster model in a form specific to one piece of software (FORTRAN). Once the FORTRAN progam is compiled, then we reach the final stage of the modelling process – actual files on a particular type of computer, or the Physical Computational Model. Although these are important, they are the preserve of software designers, and need not concern us here.

In the Worboys scheme, the term Logical Computational Model is used to refer to the way in which the vector and raster models can be represented in software. The term data structure is also commonly used in the GIS literature for this same stage in the modelling process, especially when discussed in conjunction with the design of algorithms to process the data and this usage will be adopted for the rest of this book.

1.3   STRUCTURE OF THE BOOK

The first three sections of this chapter have provided a brief overview of the topics which will be explored in more detail in this book. The book has two main aims. The first is to provide some insight into the workings of a GIS for those with no previous knowledge in this area. This is not strictly necessary for many users of GIS. Indeed the authors of a recent GIS textbook (Longley *et al.*, 2000) have argued that previous textbooks in this area have concentrated too much on the technical details of the software, at the expense of some of the other issues which impact on the daily use of spatial information. I would certainly agree that it is not necessary for every GIS user to have a full mastery of the programming which underlies a particular GIS package. However, I would argue that for many people, some knowledge of what is going on inside their GIS system is likely to be interesting and may be useful. A good analogy here is the motor car. It is not necessary to understand the workings of the internal combustion engine to drive a car, but many people find that a little knowledge of the principles is helpful.

The second aim of the book is to provide an introduction to some of the key concepts and ideas from Computer Science, in the development of data structures and algorithms. The intention is to provide sufficient background knowledge for those who wish to read more widely in this field by giving a flavour of the way that Computer Scientists tackle problems and by introducing some of the more important terminology. Terms will be defined as they are introduced, but there is also a separate glossary at the end of the book.

The structure of the book is based on the different types of spatial data which are commonly handled in GIS rather than being structured around key concepts in data handling. Computer Science concepts are therefore introduced in a relatively informal manner, in a context in which they are particularly relevant. For example, the idea of a data structure called a tree is introduced in the context of handling raster data using quadtrees, although it is equally applicable to sorting points into sequence or handling network data. One consequence of this is that the book is best tackled sequentially, at least at a first reading, because later chapters will assume a familiarity with material covered earlier in the book. The book begins by considering some of the data structures and algorithms used to handle vector data. One of the key issues in the design of both data structures and algorithms is that they should be efficient. Data structures should not use more memory or filespace than necessary, and algorithms should run quickly. As well as describing vector data structures and algorithms, Chapters 2–4 will provide you with enough background to understand the way in which efficiency is assessed in practice, and this is the subject of Chapter 5. The notation and ideas which are introduced are then used throughout the remainder of the book. Chapters 6 and 7 describe the data structures and algorithms used for raster data. Chapter 8 then discusses how large spatial databases can be indexed, so that individual features can be accessed efficiently, a topic which is equally relevant to both vector and raster databases. Chapters 9–11 consider geographical phenomena which can be modelled using both vector and raster, and compares the algorithms which are used in each case. Many geographical phenomena can be modelled as surfaces, which are considered in Chapters 9 and 10. The most obvious is the surface of the Earth itself, but the main characteristic of a surface is that there is a value for 'elevation' at every location, and so a surface can be used to model other phenomena, such as temperature or soil pH, which have a value at every point in space. Although vector and raster data structures exist for surfaces, the raster ones are more widely used, and some of the reasons for this are explored. Chapter 11 considers networks, which in some ways are the opposite of surfaces in that they are more naturally represented using vector data structures. After a discussion of why this is the case, a range of common network algorithms are explored in detail.

In order to improve the readability of the book, references to the literature have been kept to a minimum within the main body of the text. Each chapter has a further reading section at the end, which indicates the sources which have been used, and also some further material which extends what has been covered. For all the topics covered in this book there is a wealth of material on the World Wide Web. Links to some of the most useful pages have been provided in the Further Reading sections, but this only scratches the surface of what is available. Using a decent search engine on almost any of the topics covered here will yield a useful link including online sets of course notes, interactive demos and academic papers.

Computer Science is strongly rooted in mathematics. Indeed, many of the early breakthroughs in the development of machines for computation were made by mathematicians. Some mathematical notation is therefore inevitable in the book, although this is at an elementary level and should not prove a problem to anyone with the normal level of secondary school mathematics. The introductory sections have introduced the idea of the algorithm and described one using normal English. In practice, algorithms are written using a computer language and we have had a short example of FORTRAN. Although this is probably comprehensible to most people, even if they do not know FORTRAN, there are lots of details in any given language which are important for the practical execution of programs, but which get in the way when explaining principles. For this reason, algorithms are commonly described using what is called 'pseudo-code', which is a sort of shorthand computer language used to describe the main sequence of events in a program, without giving all the details. As an example, here is a program to read in the landuse array and count how many pixels have the value 10 written in FORTRAN and pseudo-code:

`      INTEGER LANDUSE(100,100)`
`C`
`      OPEN(UNIT=1,`
`    + FILE='LANDUSE.DAT')`
`      READ(1,*) LANDUSE`
`C`
`      N10 = 0`
`      DO 100 I = 1,100`
`      DO 100 J = 1,100`
`100  IF (LANDUSE(I,J).EQ.10)`
`    + N10 = N10 + 1`
`C`
`      PRINT *, N10,`
`    + 'Pixels of landuse 10'`
`      STOP`
`      END`
	`Array LANDUSE`
	`For each Row`
	`  For each Column`
	`   If (LANDUSE eq 10)`
	`    n = n + 1`
	`Print n`

FORTRAN	Pseudo-code

The FORTRAN version contains quite a lot of 'housekeeping' detail – the file containing the data must be named, the array size given, the variable to hold the answer set to 0 and so on. In contrast, the pseudo-code example simply focuses on the main structure of the algorithm; use an array for the data and process each row, a column at a time, keeping a tally of the number of times 10 is encountered. Pseudo-code can often help support the written description of an algorithm and will be used in this way in the remainder of the book.

There is no such thing as 'standard' pseudo-code, in the way that there are standard definitions of languages such as C or FORTRAN. Different authors use different notation, to suit their purpose. The only rules are that:

- The pseudo-code should convey the operation of the algorithm being described clearly.
- It should be possible for a programmer to implement the algorithm in any suitable language. This means the pseudo-code should not rely on features which are only available in one language.

It is quite normal to find descriptive English sentences as part of pseudo-code, if this is the clearest way of describing what needs to be done. However, pseudo-code tends to resemble a simplified form of programming language and will present no problem to anyone with a knowledge of any modern programming language. For those with no programming experience, the next section will give a brief introduction to some of the main features common to most programming languages, and to the syntax of the pseudo-code which will be used in this book.

1.4   PSEUDO-CODE

The easiest way to describe pseudo-code is to show how it can be used to describe a simple algorithm. The example which has been chosen is an algorithm for finding prime numbers. These are numbers which can only be divided by themselves and by 1. For instance, 4 is not prime, because it can be divided by 2, but 3 is prime. The number 1 itself is a special case and is considered not to be a prime number. A simple method of finding primes is called the 'Sieve of Eratosthenes' after the early Greek scientist who first described it. We know that any number which is divisible by 2 cannot be a prime. Therefore, if we start at 2, every second number is not a prime and can be deleted from consideration. Starting from 3, every third number cannot be a prime and can also be deleted. We already know that 4 is not prime, because it was deleted when we considered mutliples of 2. We also know that multiples of 4 will already have been dealt with in the same way. We therefore skip to the next number which has not been deleted from our list, which is 5. In theory, this method would identify all the primes up to infinity. In practice, we can only do the calculations up to a defined limit. If we decide to find all primes up to 100, we only need to apply our method starting with numbers up to 10, the square root of 100. There is no need to consider numbers which are multiples of 11. 11 multiplied by more than 11 will be greater than 100, because both numbers are bigger than the square root of 100. 11 mutliplied by anything less than 11 has already been dealt with.

Let us start by writing out this algorithm as a series of steps:

```
1   Make a list of numbers from 1 to 100.
2   Mark number 1 since it is not a prime.
3   Set k=1.
4   Until k equals or exceeds 10 do steps 6 to 8
```

```
      Find the next unmarked number in the list after k.
5     Call it m.
6     Starting from m, mark every mth number in the list.
7     Set k=m.
```

Note that this description of the algorithm is effectively already in pseudo-code, because it contains enough information for a programmer to write a computer program to run it. However, let us convert it into a form of pseudo-code which more closely resembles an actual programming language. The format used is based on the one used by Cormen *et al.* (1990).

First, we are going to need a set of 'boxes', numbered from 1 to 100 representing the numbers. In programming this is called an array, and we will represent it as follows:

```
NUMBERS[1..100]
```

Each box will have to be able to hold a label indicating whether it has been marked or not. The simplest method is to put a NO in every box at the start, and label those which have been marked with a YES. To put a value in one box, we use this syntax:

```
NUMBERS[1]=NO
```

To set all the elements of NUMBERS to NO, we can use the idea of a loop, which repeats an action a set number of times:

```
for i=1 to 100
  NUMBERS[i]=NO
```

Here i is a variable which can hold any single item of information – a number or a character string. In this case, it is initially set to 1, and then successively to 2, 3 and so on up to 100. Each time, it is used to reference the next element of the NUMBERS array.

Since this is pseudo-code, we probably do not need this detail, so we might equally well say:

```
Set all elements of NUMBERS to NO
```

since any programmer will know how to do this. Number 1 is a special case, so we mark it YES:

```
NUMBERS [1]=YES
```

We use another variable k and set it to 1.

```
k=1
```

The next step is to find the first number after k which is not marked. Again we can use a loop for this:

```
until NUMBERS[k]==NO or k>=rootN
  k=k+1
```

The second line of this sequence increases the current value of k by one. This is repeated until the kth element of NUMBERS is found to be NO or until k itself reaches the square root of N. The == sign in the first of these lines is the test for equality, and is deliberately different from the single equals sign. Thus, whereas

```
k=1
```

sets variable k to 1

```
k==1
```

tests whether the current value is 1 or not, but does not change it. The >= symbol indicates 'greater than or equal to' and != means 'not equal to'. We have also introduced a new variable, rootN, and we will have to calculate this in some way. The details of how this is done are irrelevant to this algorithm, so we can simply include a statement such as:

```
rootN=square_root(N)
```

Having found our starting point, we need to step through the remaining numbers marking every kth one:

```
for i=k to N step k
  NUMBERS[i]=YES
```

Here i is set to numbers between k and N in steps of k. When k has the value 2, i will take on values 2, 4, 6 and so on. Each time it is used to set the relevant element in the NUMBERS array to YES to indicate that it is divisible.

We now have almost all the elements we need for our algorithm. Let us look at the whole thing and consider the few extra things we need

```
1 /* Program to identify prime numbers
2 /* using the Sieve of Eratosthenes
3 Input:N
4 Array NUMBERS[1..N]
5 rootN=square_root(N)
6 Set NUMBERS[1..N] to NO
7 NUMBERS[1]=YES
```

```
 8 k=1
 9 until k>= rootN
10   m=find_next_prime(NUMBERS,k)
11   for j=m to N step M
12     NUMBERS[j]=YES
13   k=m
14 /* Print out primes
15 for i=1 to N
16   if NUMBERS[i]==NO print i

17 procedure find_next_prime (NUMBERS[],n)
18 until NUMBERS[n]== NO
19   n=n+1
20 return n
```

First, there are some lines starting with /*. These are comments, meant to clarify the program. Even with pseudo-code, it is sometimes helpful to include comments on what particular sections of the algorithm are trying to do. Line 3 indicates that the program requires an input, indicating the maximum number to be examined.

On line 10, we reach the point where we need to find the next number in our list which hasn't been marked as YES. Rather than include the pseudo-code for doing this directly, this has been put in a separate procedure, which is contained in lines 17–20. This has the effect of clarifying the code on line 10, by making clear what needs to be done, without describing the details. It also illustrates an important feature of real programming languages which always include some way of breaking a program into smaller elements in this way. In order to work the procedure needs to know some things, such as the current value of k, and these are all passed as what are called arguments. The first line of the procedure defines what arguments it expects – in this case, the array of numbers to be searched, the current value of k and the maximum point in the array to search. The syntax NUMBERS[ ] indicates that this argument will be an array, rather than a single variable. The final line of the procedure takes the value which has been found and returns it as the result of the whole procedure, and this is the value which will be placed in variable m on line 10 of the main program. It shoud be clear from this that on line 5 of the main program square_root(N) is also a call to a procedure, which takes a number as its argument, and returns the square root as the result. The code for this procedure is not given in detail, because it is of no relevance, and because many languages provide a square root procedure as standard. This simple program has covered many of the main features of programming languages:

- The storage of information in variables.
- The storage of lists of information in arrays.

- The idea of looping to repeat and action, with tests to indicate when to end the loop.
- The use of procedures to break a task down into smaller pieces.

It should provide sufficient background for you to understand the pseudo-code examples used in the rest of the book.

FURTHER READING

In 1990, the American National Centre for Geographic Information and Analysis published the Core Curriculum for GIS. This was a series of lecture notes, with associated reading and exercises which included, among other things, lectures on data structures and algorithms. The entire original set of materials is now available on the World Wide Web at the University of British Columbia web site: http://www.geog.ubc.ca/courses/klink/gis.notes/ncgia/toc.html. An updated version of the curriculum was also produced online (http://www.ncgia.ucsb.edu/pubs/core.html), and although it is incomplete, it still contains much useful material.

Two GIS textbooks (Worboys 1995; Jones 1997) have been produced by computer scientists, and therefore contain somewhat more technical material than some of the other texts currently available. Both contain details of vector and raster data structures, and of key algorithms. Worboys (1995) is generally the more technical of the two, and has a considerable amount on databases, while Jones (1997) has a focus on cartographic issues, including generalization. The book by Burrough and McDonnell (1998) also contains some details of data structures. The volume edited by Van Kreveld *et al.* (1997), while not a textbook, contains a series of good review papers covering different aspects of GIS data structures and algorithms. Maguire *et al.* (1991) and Longley *et al.* (2000) (affectionately known as the first and second editions of the Big Book) also contain a good deal of useful material.

Several introductory texts exist on the subject of Computational Geometry, which deals with computer algorithms to solve geometrical problems, such as solving mazes. The most accessible for a non-technical audience, and one with a strong emphasis on GIS applications of computational geometry, is De Berg *et al.* (1997). Excellent guides to Computational geometry resources on the web are maintained by Godfried Tousaint (http://www-cgrl.cs.mcgill.ca/~godfried/teaching/cg-web.html) and Jeff Erickson (http://compgeom.cs.uiuc.edu/~jeffe/compgeom/)

The American National Institute of Standards and technology maintain an excellent dictionary of algorithms and computer science terms on the web at http://hissa.ncsl.nist.gov/~black/DADS/terms.html. As well as providing dictionary definitions of terms, the site often has links to other pages, which provide lengthier explanations and often give examples. The FOLDOC dictionary of computing (http://foldoc.doc.ic.ac.uk/foldoc/index.html) is also useful, but has less in the way of explanation.

Those wishing to access the academic computer science literature would do well to start by using the searchable, online database of the scientific literature in this area maintained by the NEC Institute (http://researchindex.org/). As well as providing abstracts, many articles are available in full, in PDF format.

The general issue of data modelling, and the importance of good algorithm design in GIS is covered by Smith *et al.* (1987). Fisher's short paper (1997) raises some interesting issues concering what is actually represented in a pixel in a raster GIS, and some of these points are expanded upon in guest editorial in the International Journal of GIS by Wise (2000).

The sieve of Eratosthenes algorithm is taken from the description on the following web page: http://www.math.utah.edu/~alfeld/Eratosthenes.html. The web page itself has an animated version of the algorithm in action.

2   Vector data structures

In Chapter 1, we have seen that in the vector data model spatial entities are stored as either points, line or areas. In this chapter, we will take a look at the key features of some of the many data structures which have been developed to store vector data. In doing so, we will discover that ideas from mathematics have been very important in the development of these data structures.

2.1   STORING POINTS AND LINES

Figure 2.1 shows a simple vector layer containing examples of all three vector data types: points (e.g. the houses), lines (e.g. the roads and rivers) and areas (e.g. the woodlands, fields). The locational information for these features is stored using geographical coordinates but how is this information actually stored on the computer?

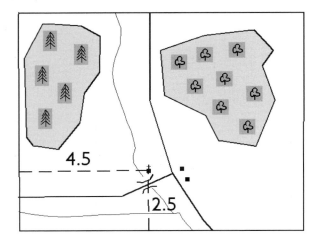

Figure 2.1 Imaginary topographic map.

Table 2.1 Coordinate data for points

Building 1	4.5	2.5
Building 2	5.8	2.9
Building 3	6.0	2.2

Let us start with points, since these are the simplest features. Consider the church near the bridge. We need to be able to store the position of this building and to do this we can measure its distance from the left hand corner of the map both horizontally and vertically as shown. This will give us a pair of figures – 4.5, 2.5. If we do the same for all three buildings, we can produce a small table (Table 2.1).

This table is clearly very suitable for storage on the computer – it could be held in an Excel spreadsheet for example or as a table in a dBase database.

Using this information we can draw a new version of our map by taking a piece of graph paper and plotting the positions of the three points. What is more, we can measure the distance between any two points. So by storing the location of the points as a pair of coordinates, we can do three things:

1 Store the locational data from the map on a computer.
2 Use this information to reproduce the map.
3 Make simple calculations from the data.

In practice, measuring the coordinates from the origin of the map is not very useful, especially if the area we are interested in covers more than one map sheet. Almost all maps have a grid on them which allows us to read off the position of points in some more useful coordinate system. Many European countries for example use the UTM (Universal Transverse Mercator) system which gives a standard way of representing positions for any point in the country. The principle is still the same however – the vector representation of the point is as a pair of coordinates.

We also need to be able to store the attributes of each point, and Figure 2.2 indicates one possible way of doing this – using extra columns in the table.

	X coordinate	*Y* coordinate	Feature Code	Building material	Name
Building 1	4.5	2.5	Church	Stone	St. Saviour's
Building 2	5.8	2.9	House	Brick	1, The Green
Building 3	6.0	2.2	House	Stone	The Larches

Figure 2.2 Attribute data for points.

The first new column is what is known as a feature code. This indicates what category a feature falls into – in the figure each building has been identified as being either a church or a house. This information could then be used when redrawing our map to determine what sort of symbol to use to represent each feature – a small black square for a house, a special symbol for a church for example.

The other columns are used to record information which is specific to each building, such as its name (if appropriate) and building material.

Now let us see how we can extend this idea to enable us to store the same data for lines and areas. Consider the road coming down from the top of the map between the two woodlands. Large parts of it are relatively straight, and so we can approximate its course by a series of short straight lines. To store the location of the lines, we simply need the coordinates of the points between each straight section, and so this gives us the following as our representation of the road.

4.5	10.0
4.5	5.7
5.5	2.5
6.5	0.3
6.8	0.0

Figure 2.3 Coordinate data for part of a line.

Using this set of numbers we can redraw our road in the same way as we did with the points – in this case we would plot the position of each point on a sheet of graph paper, and join successive points with a straight line. We can calculate the length of any of the straight sections simply by calculating the distance between the points at each end – adding all these up gives us an estimate of the length of the line.

With a curved line, such as a stream, this representation by a series of straight sections will only be an approximation of course as shown in Figure 2.4.

The only way to improve the match between the real line and our series of straight segments is to add extra points along the line. The more points we add, the closer the match, but of course each extra point means more data to be stored and hence larger files on the computer.

Again, the basic table of *XY* coordinates for a line is a relatively simple file which could be stored in a spreadsheet or a database package. However, we also want to store attributes for this line, and this is when we begin to run into difficulties.

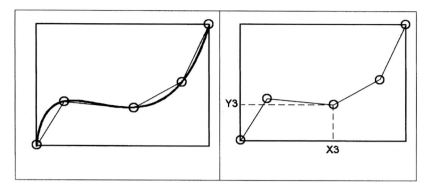

Figure 2.4 Approximating a curved line with a series of straight segments. Left: original line in black, with positions of digitized points connected by staight lines in grey. Right: digitized representation of the original line, showing X and Y coordinates for third point.

Imagine that we wish to store the following information about our roads.

Name	Surface quality	Peak traffic flow
A23	Fair	1000
A231	Good	340

Figure 2.5 Attribute data for lines.

This is a very simple table. The problem comes when we try and combine both sets of information into one table. For each road in our table, its course is represented by a large number of XY coordinates rather than the single set we had for a point feature. We could try and simply add the X and Y columns on as we did with the points (Figure 2.6).

This means that each feature is now represented by a different number of rows, depending on how many XY coordinate pairs we have. To keep each feature in a single row, we could add more columns to each row as shown in Figure 2.7.

However, now we have different numbers of columns for each feature. We could add an extra column to each row indicating the number of XY coordinates used to represent that road. However whichever way we do it, we still have a rather messy table to deal with, compared with the neat simplicity of the table for point data. As an added difficulty, many of the database systems commonly used today will not allow tables where each row has a different number of columns, or where a single feature is

Name	Surface quality	Peak Traffic Flow	*X* coordinate	*Y* coordinate
A23	Fair	1000	4.5	10.0
			4.5	5.7
			5.5	2.5
			6.5	0.3
			6.8	0.0
A231	Good	340	0.0	1.5
			3.6	1.5
			5.5	2.5

Figure 2.6 Adding locational information to attribute table for lines – rows containing data for A23 are shaded light grey.

Name	Surface quality	Peak traffic flow	X_1	Y_1	X_2	Y_2	X_3	Y_3	X_4	Y_4	X_5	Y_5
A23	Fair	1000	4.5	10.0	4.5	5.7	5.5	2.5	6.5	0.3	6.8	0.0
A231	Good	340	0.0	1.5	3.6	1.5	5.5	2.5				

Figure 2.7 Alternative method of adding locational information to attribute table for lines.

represented by more than one row – this is certainly true of the relational database systems often used in GIS.

Because of this, many vector GIS systems solve the problem by storing the locational and attribute data separately. The attribute data is stored in a standard database package, but the locational data is stored using specially written software which can handle its more complicated structure. Such systems are often referred to as geo-relational, because the attributes are often held in a relational database, with the Geographical or locational data being handled separately. The best known of these systems is probably ARC/INFO, in which the attribute data is handled by a standard database package – INFO – and the locational data is handled by specially written software – ARC.

Having separated the location and the attribute data, such systems then have to make sure that they can link back together again when necessary – for example if a user selects a particular feature by querying the database, then in order to draw that feature on the screen, it will be necessary to

Building-ID	X coordinate	Y coordinate
1	4.5	2.5
2	5.8	2.9
3	6.0	2.2

Figure 2.8 Locational data for buildings.

Building-ID	Feature Code	Building material	Name
1	Church	Stone	St. Saviour's
2	House	Brick	1, The Green
3	House	Stone	The Larches

Figure 2.9 Attribute data for buildings.

retrieve its locational data. This is done by making sure that each feature has some sort of unique identifier which is stored with both the locational and attribute data.

Let us first see how this works with the point data. In the original table (Table 2.1) we had a column which simply identified each point as Building 1, Building 2 etc. Instead we will now have a column labelled Building-ID which will contain the IDentification number of each building. This number has nothing to do with any characteristic of the building in the real world, but is simply a number assigned to each building on the map. We can then split our original table into two tables (see Figures 2.8 and 2.9), one each for the locational and attribute data.

In the case of the road data, we might use the road identification number as our unique ID but this will not be a good idea if we wish to distinguish between different parts of the A23 for example, so again we will simply use a number starting at 1. Our attribute table will now be as shown in Figure 2.10.

Road-ID	Name	Surface quality	Peak traffic flow
1	A23	Fair	1000
2	A231	Good	340

Figure 2.10 Modified attribute table for roads.

2.2   STORING AREA BOUNDARIES

Now that we have covered some of the basics of storing points and lines in a GIS, let us return to the third major type of spatial feature – the area. Figure 2.11 shows a simple area of woodland, and one way to store this area is by storing the line which defines its boundary, as shown in Figure 2.11b.

We already know that we can store lines as a sequence of *X, Y* coordinate values – the only difference in this case is that the end of the line joins the start to make a closed boundary. As with the other lines, we have considered we must approximate the boundary in order to store it. The points would be stored in order, with the coordinates of the last being the same as the first – on some systems the coordinates of the last point are simply assumed to be the same as the first. Figure 2.11b is what mathematicians call a polygon – a closed shape described by a series of straight lines – and in the GIS literature the term polygon is often used to refer to areas.

As with points and lines, we will probably wish to store attributes for our areas. With points and lines we simply added a label to the point and line data stored in the database, but it makes less sense to add a label to the boundary of an area – we naturally think of attributes as being associated with the interior of an area rather than its boundary. Therefore, it is very common to store a centroid for each area, which is a point that is located inside the polygon as shown in Figure 2.11. The centroid can be defined by hand when the area is digitized, but many systems will automatically define one if this is not done. The centroid is commonly used to give a position for labels when drawing maps of area features and for this reason centroids are normally positioned near the centre of the area (as their name implies).

The use of centroids means that to store a single area in our GIS, we actually need to store two things – the line defining the boundary and the point defining the centroid. In fact, things become more complicated still

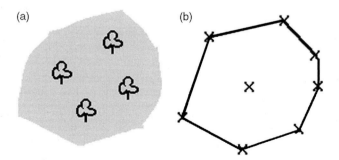

Figure 2.11 Storage of area features.

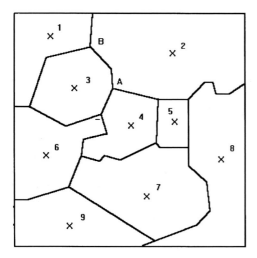

Figure 2.12 Landuse map – an example of multiple areas.

because so far we have only dealt with the simplest type of area. Figure 2.12 shows our original woodland as part of a landuse map. Rather than a single area we now have a series of areas which neighbour each other, completely covering the area of the map. This type of map is very common – other examples are soil maps, geology maps and maps showing administrative areas such as countries, counties, or districts.

Each area has a centroid, with an identifier associated with it, and this identifier is used as the link to a table containing the attributes for the areas (Figure 2.13).

We can still use the simple method of storing the areas shown on Figure 2.11 but we will run into a number of problems. If we consider polygon 2, we can see that this shares part of its boundary with our original woodland – between points A and B on the map. However, although we have already stored this part of the line in storing the woodland boundary, we have to store it again, otherwise there will be a gap in the boundary of polygon 2. If we look at the whole map we will see that the majority of the boundary lines lie between two areas, and will be stored twice in this way – the result is that we will store nearly twice as much data as necessary.

This is not the only problem. When we store the boundary, we choose a series of points along the line, and connect these by straight lines. When the same line is digitized a second time, slightly different points will be chosen, with the result shown in Figure 2.14.

This shows the part of the boundary of polygon 2 between points A and B in black and the same part of the boundary of polygon 3 in grey. Because the two lines do not coincide, there are small areas of overlap, and small gaps

ID	LANDUSE
1	Pasture
2	Pasture
3	Woodland
4	Urban
5	Arable
6	Pasture
7	Pasture
8	Woodland
9	Arable

Figure 2.13 Attributes for landuse map.

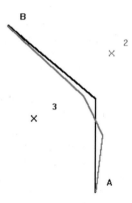

Figure 2.14 Sliver polygons as a result of digitizing the same line twice.

between the two areas. These mismatches areas are called sliver polygons, because they are usually very small and thin.

There is a third problem with this method of storing area boundaries which arises if we wish to use our data for analysis rather than simply map drawing. We may wish to produce a new GIS layer which simply shows urban and non-urban areas. To do this we have to merge together all the polygons in which the landuse is not urban – i.e. to dissolve the boundaries between them resulting in a new layer looking like the one in Figure 2.15.

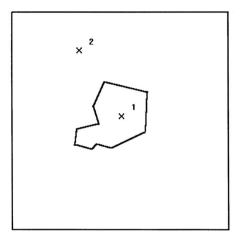

Figure 2.15 Map of urban and non-urban areas created using a polygon dissolve operation.

This operation is called a polygon dissolve, and is quite common in GIS analysis. However, it is difficult to do with the simple method of storing the area boundaries. If we consider our original woodland, we do not know which part of the boundary has an urban area on the other side (so that we will need to keep that part of the boundary) and which has non-urban landuse on the other side (so that we can drop that part of the line). In technical terms, we do not have any information about the contiguity of our polygons – which ones are neighbours to each other – and to store this information we need a different method of storing our area boundaries, which will be covered in the next section. However, it should be said that the method of storing areas described here has one great advantage, which is its simplicity. In addition, it is perfectly adequate for many of the operations we need to have in a GIS – we can store attributes for our areas, measure their areal extent and perimeter, and produce maps based on them. This is why this method of storing areas is very common in mapping packages.

2.3 STORING AREA BOUNDARIES: THE TOPOLOGICAL APPROACH

To overcome the limitations of the simple method of storing polygons, GIS systems draw on ideas first developed in a branch of mathematics called topology. The link between topology and GIS will be explored in detail in the next section, but for the moment, I will describe the way in which area data is stored in many GIS systems.

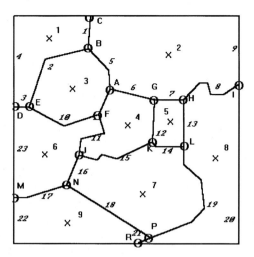

Figure 2.16 Landuse map as stored in a topological GIS.

If we look at the landuse map in Figure 2.16, we can see that each area boundary is made up of a series of line sections, which meet at junctions such as the points marked A and B. If we identify each of those junctions, and store the location of the line section joining them, we will have stored all the boundary information, but without duplication.

The junctions are called nodes, and the sections of lines between them have various names in the GIS literature – arcs, chains, segments, edges and links. Since this method of storing area boundaries is often called link-and-node, I will use the term link.

In Figure 2.16, each link has been given a number (in italic numerals) and each node identified by a letter. For each link, we can store the location of the end points (the nodes) and of a series of positions along the line recording its location. However, we also need to know which link belongs to which areas, remembering of course that most links form part of the boundary of two polygons. This is done by storing for each link the identifier of the polygons on either side. For this purpose, the links must have a direction so that we know which side of the line a polygon is on – so for link 5 for example we might store (Figure 2.17).

Link ID	Start Node	End Node	Left Poly	Right Poly	X_1	Y_1	X_2	Y_2	etc.	X_n	Y_n
5	A	B	3	2							

Figure 2.17 Information stored for a link in a link and node data structure.

The direction of the link is not important – we could equally well store this link as running from B to A with polygon 2 on the left and 3 on the right.

The same information is stored for every link on the map. Links such as number 4 which form part of the border of the map only have a polygon on one side – here it is customary to assign a number to the area 'outside' the map, so that link 4 might be recorded as having area 1 on one side and area 0 on the other.

Since we know which areas lie on either side of each link, we can use this information to construct the boundary of any area. For example, to construct the boundary of polygon 3 (our original woodland), we go through the complete set of links identifying all those with 3 to the left or right. We then attempt to join them together into a closed boundary by matching the node identifiers. For example, if we start with link 5, this ends at node B – we then need a link which joins it at node B. Depending on the direction it has been digitized, this may start or end with node B so we must check for both. We would continue this process until we arrived back at node A. In Figure 2.16, this process would fail because one of the links bordering polygon 3 has not been given an identifier and so will not be stored in the database. This illustrates one of the strengths of the link and node structure – it can be used to check for errors in the data (which is how this structure originated).

The structure also solves our other problems. First, each link is only stored once, thus saving the problem of duplicating data. Second, we now have information about the relationship between areas, which can be useful for analysis. For example, if we wish to expand the urban area in polygon 4, it is simple to identify which land parcels border it, and which might therefore be considered as potential development sites.

The same link and node structure can also be used for line data, where it is the connections between lines at the nodes which is the important element – the left/right area identifiers are generally ignored. However, knowing about connections between lines means that we can trace routes through networks for example, a topic which will be covered in Chapter 11.

The key to the link and node method of storing area boundaries is in the information which describes the relationships between objects on the map – the left/right identifiers and the to/from node identifiers for each link. Because of the origin of this structure in the mathematical subject of topology, this information is often described, somewhat loosely, as topological data. The X, Y coordinate pairs giving the location of points along the line are collectively known as the geometrical data. The two are shown together in the table for link 5 above, but in some systems they are actually stored separately. This means that in many GIS systems, the storage of area data is rather complex since each area requires one centroid and one or more links describing the boundary. In turn, each link will require one entry for the topological data plus one for the geometrical data. To understand how this information is used within a GIS, we need to go back to the maths classroom once more and understand a little elementary topology!

2.4   SO WHAT IS TOPOLOGY?

The study of relationships such as contiguity (whether objects are neighbours or not) is part of the mathematical subject of topology, which is concerned with those characteristics of objects which do not change when the object is deformed. For example, imagine the landuse map shown in Figure 2.16 printed on a sheet of thin rubber. If the rubber were stretched, then some properties of the areas would change, such as their size and shape. However, no amount of stretching could make polygon 3 border polygon 7 – this would involve cutting the sheet or folding it over on itself. Hence, the connection (or lack of it) between areas is a topological property of the map. Containment is another example of a topological property, since no amount of stretching will move centroid 3 outside its polygon. One of the earliest people to study such properties was the Swiss mathematician Leonhard Euler (pronounced 'Oiler') and one of the classic problems he studied, the Konigsberg bridge problem, has a direct relevance to the use of topological ideas in GIS. In the town of Konigsberg, there was an island in the Pregel River, which was connected to the banks of the river by seven bridges as shown in Figure 2.18.

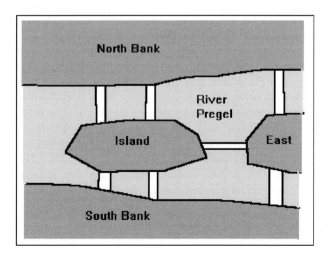

Figure 2.18 The Konigsberg bridge problem.

The local people believed that it was impossible to plan a route which started and ended in the same place but crossed every bridge only once. However, nobody could prove whether this was in fact correct. Euler realized that the problem had nothing to do with the distances or directions involved, but depended solely on the connections between places. He reformulated the problem, by representing each of the land masses as points, and the connections between them as lines (Figure 2.19).

This representation is called a graph by mathematicians. The key to the problem is the number of lines which meet at any given vertex of the graph – if this is an even number you can reach that vertex and leave by a different line. If it is an odd number, then eventually the pairs of entry/exit lines will be used up and there will be only one unused line joined to that vertex – i.e. you can visit the vertex but can't leave again without using a line for the second time. Therefore, it is only possible to make a round trip walk if all vertices have an even numbers of lines, or if there are just two vertices at which an odd number of lines meet (which will have to be the start and end points of the route). In the case of the Konigsberg bridges neither condition is satisfied, proving that the round trip cannot be made and that the locals were right.

Another mathematician, Henri Poincaré, realized that graph theory could be applied to maps in general, and his ideas were used by staff at the US Bureau of the Census to help in processing the data for the 1980 census. A key part of processing the census data was in handling the map of the street network of the United States of America. This indicated where each address was located and which block it fell in. In order to automate the processing of census data, it was necessary to have an accurate database which indicated which block each address was in. Compiling such a database was an enormous task in which errors were bound to be made, and so some way was needed for checking the accuracy of the results.

The map in Figure 2.20 is a fictitious example of part of the street network in an American city. It can be seen that each block is surrounded

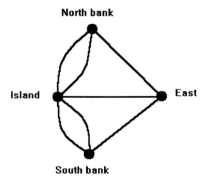

Figure 2.19 Graph of the Konigsberg bridge problem.

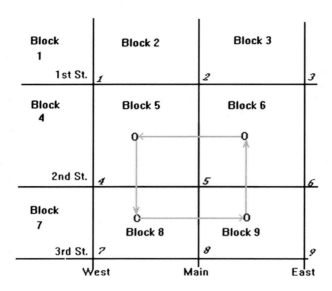

Figure 2.20 Fictitous city blocks illustrating Poincarés dual graph model of maps.

by sections of street which meet at junctions. If we treat each street inter-section as a vertex, we can regard the street network as a mathematical graph. What is more, if we consider the part of the graph which surrounds an individual block (e.g. Block 5 in the above map) it is clear that this will form a single connected circuit.

However, we can also use a graph to represent the relationship between the blocks. First, we represent each block by a single point as shown with blocks 5, 6, 8 and 9 in Figure 2.20. The points are then joined by lines if they lie on either side of a street section and we have a graph similar to the one constructed by Euler for the Konigsberg bridge problem. If we take the portion of this graph which surrounds a single street intersection (like the example shown in grey in Figure 2.20 which surrounds node 5), then this should form a single connected circuit as with the first graph.

We can therefore construct two graphs – one based on the streets sur-rounding a block, one on the blocks surrounding a street intersection, and it was this model of maps as paired graphs which came from the work of Poincaré. Mathematically, the two graphs are exactly the same, since both will consist of a single closed loop. If we can automatically create these graphs from our street map, and check that they do form closed loops we will have a way of checking the data. This is exactly what the staff at the Bureau of the Census managed to do when they developed a data structure called DIME, which will be explained in the next section.

2.5 AND HOW DOES IT HELP? THE EXAMPLE OF DIME

So far we have considered these graphs by drawing them on the original map. However, computers cannot 'see' maps, so we must devise a method of storing the data in a format which allows the computer to construct and test the graph around any block or junction. The system which was devised was called DIME (Dual Independent Map Encoding) and was based upon a data structure in which one record was created for each street segment – the part of a street between two junctions. For the map in Figure 2.20, a portion of the DIME file might be as shown in Figure 2.21.

If we look at block 5 in Figure 2.20, it is very simple for us to see that it is surrounded by four connected street segments. In order to check this on the computer using the data structure we first find those records in the DIME file which have 5 on either their left or right. Since the DIME file is simply a table, this could be done using a standard database query, or by writing a program to read each record and select those in which the block left or block right value was 5. In this case, we will find segments 1, 3, 7 and 9. We then need to check that these segments form a closed loop, and this is most easily done if they all go in the same direction around the block. If they all have block 5 on their right hand side, this means they will form a clockwise loop. At the moment, only segments 1 and 7 have 5 as their right

Seg#	From	To	Block Left	Block Right	Street Name
1	1	2	2	5	1st
2	2	3	3	6	1st
3	4	5	5	8	2nd
4	5	6	6	9	2nd
5	7	8	8	11	3rd
6	8	9	9	12	3rd
7	4	1	4	5	West
8	7	4	7	8	West
9	5	2	5	6	Main
10	8	5	8	9	Main

Figure 2.21 DIME data structure for fictitious city map.

Seg#	From	To	Block Left	Block Right	Street Name
1	1	2	2	5	1st
3	5	4	8	5	2nd
7	4	1	4	5	West
9	2	5	6	5	Main

Figure 2.22 Records from DIME file relating to block 5. All records have been modified so that block 5 is on the right hand side of the street.

hand block – however to change the direction of the other two all we need to do is switch the left and right blocks and the to and from nodes to produce the records shown in Figure 2.22.

We can now start at any segment and try and trace round the loop. We start at segment 1, noting that our start point is node 1. The end of this segment is node 2, so we look for a segment which starts at node 2. We find segment 9, which ends in node 5. This leads us to segment 3 which ends in node 4, and this leads us to segment 7, which ends in node 1, our starting point. If for any reason we can't complete this loop, we know there is an error in the data, such as a segment completely missed out, or one in which the block numbers or node numbers were wrong. For example, if segment 2 had 5 as its right block instead of 6 this would create three errors:

- Block 3 would not be correct because of the 'missing' segment.
- Block 6 would not be correct because of the 'missing' segment.
- Block 5 would close but would have a segment unused.

This checking process can also be carried out using the graph based around the street junctions. If we consider junction 5, then we can identify the segments which meet at this point because they will have 5 as either their start or end node (Figure 2.23).

We then adjust these so that they all end at node 5 as shown in Figure 2.24. Now if we start at a segment, then the block on the right of that segment must be the left hand block of one other segment, which in turn will share its right hand block with one other segment, until we work our way around to the starting point.

In both the block and junction checking, we are using the left/right and from/to information to trace around the topological graph. Since we know

Seg#	From	To	Block Left	Block Right	Street Name
3	4	5	5	8	2nd
4	5	6	6	9	2nd
9	5	2	5	6	Main
10	8	5	8	9	Main

Figure 2.23 Records from DIME file relating to junction 5.

from mathematical theory that there must be one closed graph in each case, if we fail to find this, we know we have an error in the data.

Notice that the original DIME file did not contain any geographical coordinates. The geographical referencing was done via addresses, since the original file had extra fields which have not been shown so far, which indicated the range of addresses on either side of the segment. This allowed any address to be matched with its block number (and census tract number), and also allowed summary statistics to be produced for any block by aggregating data for all the addresses (Figure 2.25).

We have now seen how ideas from topology lead to the DIME data structure. Since this was intended simply for handling data for streets, the segments in the DIME file were all straight – if a street curved, it was simply broken up into segments. In order to develop this into a more general purpose data structure, it was necessary to allow the lines between the nodes

Seg#	From	To	Block Left	Block Right	Street Name
3	4	5	5	8	2nd
4	6	5	9	6	2nd
9	2	5	6	5	Main
10	8	5	8	9	Main

Figure 2.24 Records from DIME file relating to junction 5 modified so that the street ends at junction 5.

Seg#	From	To	Block Left	Block Right	Street Name	Left Address Low	Left Address High	Right Address Low	Right Address High
1	1	2	2	5	1st	12	24	13	25

Figure 2.25 Storage of address information in the DIME data structure.

to take on any shape, as described by a set of XY coordinates – in this way we reach the link and node data structure which was described earlier.

FURTHER READING

A number of authors have discussed the data structures used to handle vector data. Peucker and Chrisman (1975) is a standard reference here, as are the works by Peuquet on both vector and raster (1984, 1981a,b). More recent work includes a discussion by Keating *et al.* (1987). Nagy (1980) has an interesting discussion of how to store point data. Because points are so simple, this would seem to be trivial, but if the intention is to make it easy to answer queries such as 'find all points within 10 km of a given location', then he shows that it is not at all easy to determine how best to structure the data.

The difficulties of storing spatial data in standard relational databases have been covered only briefly here. A more extensive discussion of the problems, and why georelational GIS systems were designed, is provided by Healey (1991). Some of the limitations have now been removed and some commercial database products have the capability of handling spatial data. Samet and Aref (1995) provide a good introduction to some of the developments in this field.

The background to the DIME data structure, and its successor, called TIGER, are given in two papers in Peuquet and Marble (1990). The paper by White (1984) is a very readable summary of the importance of topological structuring in handling vector data.

Worboys (1995) and Jones (1997) both have good sections on vector data structures, giving detailed examples which go beyond the material covered in this book.- The NCGIA core curriculum (http://www.geog.ubc.ca/courses/klink/gis.notes/ncgia/toc.html) Unit 30 covers the structures described here, while Unit 31 describes chain codes, which are an alternative data which are a very compact way of representing vector lines. Nievergelt and Widmayer (1997) provide an excellent review of why data structures are needed and why specialized ones have had to be developed for spatial data.

The algorithms and data structures used for vector spatial data have much in common with those used in vector computer graphics. For example, to window in on a detailed line drawing, such as an architect's plan, it is necessary to decide which lines lie within the window, which outside and which cross the window boundary – in other words a classic line intersection problem. The standard reference work on computer graphics is Foley *et al.* (1990) but Bowyer and Woodwark (1983), is an excellent, simple introduction to some of the fundamental algorithms. The FAQ file from the comp.graphics.algorithms newsgroup also has some useful information and is archived at http://www.faqs.org/faqs/graphics/algorithms-faq/.

3 Vector algorithms for lines

Because points, lines and areas are so different from one another, the type of queries they are used for and hence the algorithms developed for them, are sometimes different. Some queries are just not meaningful for all vector types. For example, polygons can contain points but points cannot contain polygons, so a point in polygon test is useful but a polygon in point test is not. Even when queries are meaningful for all types, they may need to work rather differently. For instance, it is possible to measure the distance between any two features. If both are points, this is simple, but if they are areas, then there are several possible distances: the distance between the nearest points on the boundaries, the distance between the centroids.

The algorithms which are described in this chapter and the next one have been chosen because they are among the most fundamental vector algorithms. Not only are they useful in their own right, but they often underpin other more complicated algorithms. They also illustrate some important characteristics of good algorithm design, and the importance of the topological data structure introduced in Chapter 2.

3.1 SIMPLE LINE INTERSECTION ALGORITHM

We have already seen (in Section 2.1) how a vector GIS can store a line feature by storing the coordinates of a series of points along the line. This section and the next two will deal with how we can use this representation to decide whether two lines intersect or not. This sort of question may be useful in its own right – for example to see whether a proposed new route crosses any existing routes or rivers, which would necessitate building bridges. Working out whether lines intersect is also a key part of many other GIS operations, such as polygon overlay and buffering.

The problem then is to decide whether two lines, such as those shown in Figure 3.1 cross each other. This might seem a trivial question to answer – most people reading this could give the correct answer instantaneously and with no conscious thought. However, as we shall see, to get a computer to give the same answer is not at all an easy matter.

Let us start by simplifying the problem. We know that each of the lines in Figure 3.1 is represented by a series of short straight segments between pairs of points. Let us first see how the computer decides whether two of these straight segments intersect or not, by considering the example of the two segments in Figure 3.2.

Again this is a trivial question for a human being to answer but far more difficult for a computer. The first reason for this is that we can see both lines whereas all the computer can 'see' is the eight numbers representing the start and end points of each line segment as shown in Figure 3.3.

The question which the computer has to answer therefore is 'Do the two line segments with these start and end coordinates intersect?'. The answer to this question is much less obvious even to a human observer, and clearly we need a method to work it out. What we need is a set of steps we can take which will allow us to take these eight coordinate values and work out whether the line segments they represent intersect – in other words an

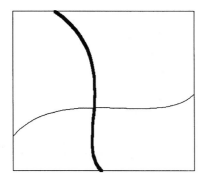

Figure 3.1 Line intersection problem.

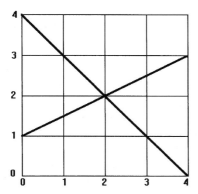

Figure 3.2 Testing for the intersection of two line segment intersections.

		X	Y
Line 1 :	Start	0	1
	End	4	3
Line 2:	Start	0	4
	End	4	0

Figure 3.3 Coordinates of line segments.

algorithm. For our first attempt at an algorithm we will need to invoke the aid of some elementary mathematics.

For simplicity, the term line will be used to refer to the line segments. We will return to the issue of lines made up of a series of segments in due course. One equation commonly used to represent straight lines is:

$$y = a + b \cdot x$$

(The dot between b and x means 'multiply'). For the two line segments in Figure 3.2, the equations are

Line 1: $Y = 1 + 0.5X$
Line 2: $Y = 4 - X$

This means that for any value of X we can calculate Y – that is if we know how far along the X axis we are, we can also work out how high up the Y axis the line is at that point. For line 1, therefore, when X is 3, Y will be $1 + 3 \times 0.5$ which is 2.5.

The important point about this is that if the two lines cross, there will be a point where the X and Y values for the lines are the same i.e. the value of Y on line 1 will be equal to Y on line 2. In mathematical notation:

$$Y_1 = Y_2 \quad \text{and} \quad X_1 = X_2$$

If we call these common values X and Y then our two equations for the lines become

$$Y = 1 + 0.5X$$
$$Y = 4 - X$$

which means that

$$1 + 0.5X = 4 - X$$

There is now only one quantity in this equation which we don't know i.e. X, and we can rearrange to give the following

$$0.5X + X = 4 - 1$$

$$1.5X = 3$$

$$X = \frac{3}{1.5}$$

$$X = 2$$

This means the two lines cross at the point where $X = 2$ (which agrees with the diagram, which is reassuring!). To find the Y coordinate for this point we can use either of our equations for our lines:

$$Y_1 = 1 + 0.5X$$

$$Y_1 = 1 + 0.5 \cdot 2$$

$$Y_1 = 2$$

Therefore, the two lines cross at the point $(2, 2)$. However, this has only showed the solution for two particular lines – in order to be an algorithm our method must be able to cope with any pair of lines. In order to do this we can generalize as follows: Given two lines with end points (XS_1, YS_1), (XE_1, YE_1) and (XS_2, YS_2), (XE_2, YE_2), find the intersection point.

1   First, we must work out the equation of the straight lines connecting each pair of coordinates. The details of how this is done are given in a separate box and this gives us values of A and B for line 1 and line 2.
2   Given values for A_1 and B_1, A_2 and B_2 we know that at the point of intersection

$$Y_1 = Y_2$$

Hence,

$$A_1 + B_1 \cdot X_1 = A_2 + B_2 \cdot X_2$$

However, we also know that where the lines meet the two X values will also be the same. Denoting this value by XP

$$A_1 + B_1 \cdot XP = A_2 + B_2 \cdot XP$$

Box – Calculating the equation of a line

Given the coordinates of any two points on a straight line, it is possible to work out the equation of the line. To understand how to do this, let us first look at what the equation of the line means. The general equation for a line is

$$Y = A + B \cdot X$$

and the equation of the line in the diagram is

$$Y = 1 + 0.5X$$

i.e. $A = 1$, and $B = 0.5$.

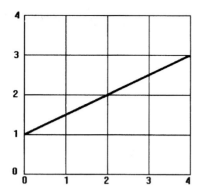

B is the slope of the line – for each unit in X it tells us how much the line rises or falls. In this example Y rises by 2 units over an X distance of 4 units – hence the slope is 2/4 or 0.5. Therefore to calculate B all we need is to look at the change in Y value between any two points and divide by the number of X units between the points.

If we have the X and Y coordinates of two points on the line, this is simply the difference in Y values divided by the difference in X or

$$B = \frac{y_2 - y_1}{x_2 - x_1}$$

In the case of this line

$$B = \frac{3 - 1}{4 - 0} = \frac{2}{4} = 0.5$$

To see what A means, consider what happens to our equation when $X = 0$. In this case

$$Y = A + B \cdot 0$$

Since anything multiplied by 0 equals 0, this means

$$Y = A$$

So A is the value of Y when X is 0. Put another way, it tells us how far up (or down) the Y axis the line is when X is 0.

In the present case we already know the answer, because one of our coordinates has a zero X value. But we can work A out from the other point. We know the Y value (3) when X is 4. We also know how Y will change by as we change X – if we reduce X by 1 unit, Y will fall by 0.5. If we reduce X by 4 units (i.e. so that it is 0) Y will fall by 4×0.5 or 2 so giving a value of 1.

In mathematical terms this means that

$$A = Y - X \cdot B$$

where X and Y can be taken from either of our points.

We can now rearrange this equation to give

$$B_1 \cdot XP - B_2 \cdot XP = A_2 - A_1$$
$$XP(B_1 - B_2) = A_2 - A_1$$

$$XP = \frac{(A_2 - A_1)}{(B_1 - B_2)}$$

This gives us our value for X at the point where the lines meet. The value for Y comes from either of the equations of our lines:

$$YP = A_1 + B_1 \cdot XP$$

We now have an algorithm which will work out where two lines intersect. However, this does not solve our problem as we shall see.

3.2   WHY THE SIMPLE LINE INTERSECTION ALGORITHM WON'T WORK: A BETTER ALGORITHM

To see what is missing from our algorithm consider the two sets of lines in Figure 3.4.

The graph on the left shows the lines used in the previous section – in the graph on the right one of the lines is somewhat shorter so that the two lines do not intersect. However, if you take the coordinates for this shorter line and use them to work out the equation for the line, they will give exactly the same result as for the case on the left: $Y = 4 - X$. The reason is that the

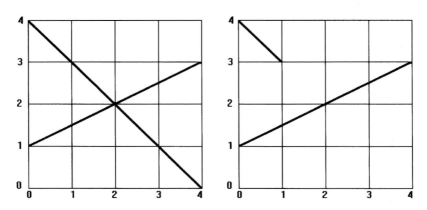

Figure 3.4 Line intersection problem – each pair of lines is defined by the same mathematical equation.

equation describes the **infinite** line which passes through the two points – it says nothing about how long the line is. Since two infinite lines will always intersect somewhere, unless they are parallel, testing to see whether they intersect or not will not tell us whether the parts of the lines we are dealing with intersect or not.

In order to finish the algorithm, we need to test to see whether the point at which the infinite lines intersect actually lies on the lines (as in example 1) or lies beyond the end of the lines (as in example 2).

To do this, we can look at the X value of the intersection point, and see whether it lies between the X values for the end of the line. From the calculations last time we know that these two lines intersect at $(2, 2)$. In example 1, line 2 runs from $X = 0$ to $X = 4$, so since 2 lies between these values, the intersection point falls on line 1. In the second example line 2 runs from $X = 0$ to $X = 1$, the intersection point lies beyond the line and so the lines don't intersect.

Note that it is not enough to test one line. If we repeat the above test for the case of line 1, this will give the same answer in both cases – the intersection point does indeed fall on line 1 in both cases. However, as we can see, in the second case the lines do not intersect – but we can only be sure that this is the case by testing both lines.

So now, do we have our algorithm? Well, yes and no! A proper algorithm must be able to deal with all situations without giving the wrong answer or causing problems, but we have so far overlooked two special cases which will trip our current algorithm up – the situation when our lines are parallel, or when either of them is vertical.

If the lines are parallel, then we know that they cannot intersect – so surely our algorithm will simply report no intersection? To see why it will not, it is useful to look at the algorithm in pseudo-code (lines starting with /* are comments to help explain what is going on).

```
/* Given end points of lines :
/*
/* Line 1:  XS1,YS1
/*          XE1,YE1
/* Line 2:  XS2,YS2
/*          XE2,YE2
/* Calculate A and B for equation of each line
/* Y=A+B*X
/* Note that the sign / is used for division
B1=(YE1−YS1)/(XE1−XS1)
A1=YS1−B1*XS1
B2=(YE2−YS2)/(XE2−XS2)
A2=YS2−B2*XS2
/* Calculate the X value of the intersection point XP
/* and the Y value YP
```

```
XP=(A2−A1)/(B1−B2)
YP=A1+B1*XP
/* Test whether point XP,YP falls on both lines
if (XP<XE1 and XP>XS1 and XP<XE2 and XP>XS2) then
   intersect=TRUE
else
   intersect=FALSE
```

To understand why parallel lines give a problem, we need to remember that in the equation

$$Y = A + B \cdot X$$

the value B represents the slope of the line. So in example 1, the line $Y = 1 + 0.5X$ has a slope of 0.5 – each step of 1 along the X axis results in a rise of the line by 0.5 units. The problem comes when we try and calculate the value of XP using the equation:

$$XP = \frac{A_2 - A_1}{B_1 - B_2}$$

If the two lines are parallel they will have the same slope, so that $B_1 - B_2$ will equal zero. In mathematics, dividing by zero conventionally gives the answer of infinity. On the computer, it causes a program to crash, usually with a message referring to 'attempt to divide by zero'. The only way to prevent this is by checking the values of B_1 and B_2 before calculating XP.

Another way to look at this is that we are actually saving work – if the two lines are parallel, they can't intersect, and we therefore know the answer to our question straight away.

A similar problem arises with vertical lines as shown in Figure 3.5. In this case, the start and end of the line will have the same X coordinate – however in calculating the values for B_1 and B_2 we need to divide by $XE_1 - XS_1$ – again when these are the same we will be trying to divide by zero. The only way to avoid this is by first testing for the situation of vertical lines. However, in this case, this does not tell us whether the lines intersect or not – as Figure 3.5 shows a vertical line may still intersect another, sloping line. However, the fact that one line is vertical makes the intersection test a little easier because it tells us the X value of the intersection point between the two lines – it must be the same as the X value of the vertical line. This means we already know XP and we can calculate YP from the equation of the sloping line.

If one of the lines is vertical however this gives us problems with the final part of the algorithm where we test to see whether the X value of the intersection point lies on each of the lines. We cannot use this test in the

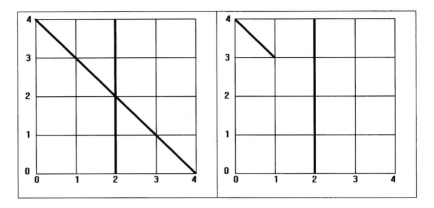

Figure 3.5 Line intersection when one line is vertical.

case of the vertical line because all the X values are the same so that the intersection point will always appear to lie on the vertical line. We can solve the problem by doing the same test in terms of the Y values rather than the X values, but then we will hit the same problem if one of the lines happens to be horizontal! The only reliable way to do the test therefore is to test both X and Y values for both lines.

At the end of all this we have an algorithm which is (almost) foolproof:

```
/* Program to test intersection of two line segments
if xs1==xe1 then
   if xs2==xe2 then
     print 'Both lines vertical'
     stop
   else                            /* First line vertical
     b2=(ye2-ys2)/(xe2-xs2)
     a2=ys2-b2*xs2
     xp=xs1
     yp=a2+b2*xp
else
     if xs2==xe2 then             /* Second line vertical
        b1=(ye1-ys1)/(xe1-xs1)
        a1=ys1-b1*xs1
        xp=xs2
        yp=a1+b1*xp
     else                         /* Neither line vertical
        b1=(ye1-ys1)/(xe1-xs1)
        b2=(ye2-ys2)/(xe2-xs2)
        a1=ys1-b1*xs1
```

```
      a2=ys2−b2*xs2
      if b1==b2 then
        print 'Lines parallel'
        stop
      else
        xp=−(a1−a2)/(b1−b2)
        yp=a1+b1*xp
/* Test whether intersection point falls on both lines
if(xs1−xp)*(xp−xe1)>=0 and
  (xs2−xp)*(xp−xe2)>=0 and
  (ys1−yp)*(yp−ye1)>=0 and
  (ys2−yp)*(yp−ye2)>=0 then
    print 'Lines cross at' xp,yp
else
    print 'Lines do not cross'
```

Two problems remain. The first is a subtle one. What result will this algorithm produce if the two line segments are exactly the same, or overlap along part of their extent? The answer is that they will probably be judged to be parallel and hence not intersect. Whether or not the lines should be regarded as intersecting or not is actually a matter of definition. The two lines could be regarded as not intersecting, or as intersecting at more than one point. If we decide that the latter is the correct interpretation, the algorithm will have to be modified to produce this result in this particular circumstance. The second problem is that because of the way the computer stores numbers, the algorithm will occasionally produce incorrect results, especially when the coordinate values are very large or very small. This is explained in more detail in Section 3.4.

3.3 DEALING WITH WIGGLY LINES

Now that we have an algorithm which will detect whether two straight line segments intersect, we can return to our original problem of detecting whether two lines, each made up of a series of segments, intersect. The simple and obvious approach to this problem is to apply our line intersection test to all pairs of segments from all lines. If any pair of segments intersect, this means the lines intersect. This will certainly work, but will take a long time – even with a fast computer. What is more a lot of these tests will be futile, because we will be testing lines which are a long way from each other, and which could not possibly intersect. As an example consider the three lines in Figure 3.6.

It is clear that lines 1 and 2 cannot possibly cross because they lie in different parts of the map, whereas 2 and 3 clearly do cross. To test whether lines 1 and 2 cross would require us to run our line intersection test 49 times (there are 7 segments in each line) so it is worth avoiding this if possible. The way this is done is to perform a much simpler test first to see whether the lines *might* cross.

In Figure 3.7, a box has been drawn around each line just neatly enclosing it – this is called the bounding box or minimum enclosing rectangle (MER). The MERs for lines 1 and 2 do not intersect – this proves that these two lines cannot intersect and that there is no need to do any further testing. So how do we construct the MER and how do we test whether they intersect?

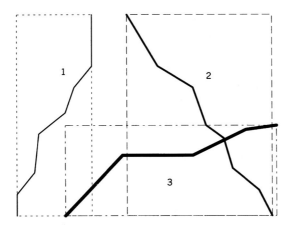

Figure 3.6 Testing for the intersection of lines.

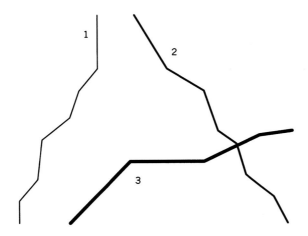

Figure 3.7 Minimum enclosing rectangle around lines.

The MER is defined by just four numbers – the minimum and maximum values of the X and Y coordinates of the line, which give the positions of the rectangle corners as shown in Figure 3.8.

So if we can find these four values, this will define our box. This is very simple – we write a program which reads through the X and Y values for all the points along the line, noting down the largest and smallest it finds. In pseudo-code:

```
1  xmin=99999999999
2  ymin=99999999999
3  xmax=-99999999999
4  ymax=-99999999999
5  for all X and Y
6    if x<xmin then xmin=X
7    if x>xmax then xmax=X
8    if y<ymin then ymin=Y
9    if y>ymax then ymax=Y
```

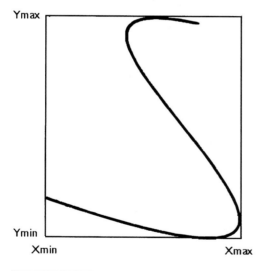

Bottom left	Xmin	Ymin
Bottom right	Xmax	Ymin
Top left	Xmin	Ymax
Top right	Xmax	Ymax

Figure 3.8 Definition of corner points of the MER.

The test for whether the MERs intersect is also simple. Consider the X values – if the smallest X value of one MER is bigger than the largest X value of the other, then the MERs cannot intersect. The same sort of logic applies to the other extremes, giving this test

```
1   if xmin1>xmax2 OR xmax1<xmin2 OR ymin1>ymax2 OR
    ymax1<ymin2 then
2      MERs CANT INTERSECT
3   else
4      MERs DO INTERSECT
```

This test would show that lines 1 and 2 cannot possibly intersect. But what about if the MERs do intersect – does this mean the lines must intersect, and it is simply a case of finding out where? As line 3 shows the answer is no – the MER for this line overlaps those of both lines 1 and 2, although the line itself only intersects with line 2.

So in the case of lines 2 and 3, our initial test has not helped, and we will still have to run our line intersection. To avoid having to run the full line intersection program for every pair of segments in lines 2 and 3 there are two things we can do. First, we can extend the idea of using the MER from the whole line to each pair of line segments. If we take the first segments in lines 2 and 3, we can imagine drawing a box around them as we did with the whole line – however, here the corners of the boxes are defined by the end points of the lines, so there is no calculation needed. Our MER test can therefore be done simply on the end points of the lines.

A further refinement is to split the line into what are called monotonic sections – sections in which the X and Y values steadily increase or decrease. In Figure 3.9, line 2 is monotonic – as X increases the Y value of line always

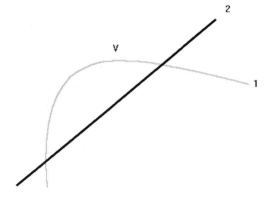

Figure 3.9 Monotonic sections of lines.

increases. However, this is not true for line 2, since to the left of the point indicated by the arrow, the *Y* values increase steadily whereas to the right they decrease steadily.

This line is therefore made up of two monotonic sections. The point about monotonic sections is that two of them can only intersect once (if they intersect at all). Looking at Figure 3.6 again, it is clear that once lines 2 and 3 have crossed they cannot cross again, unless one of them 'turns the corner' – for example if line 3 started to come down the page. This means that once an intersection has been detected between two segments within a monotonic section, the rest of the segments can be safely ignored.

At the start of Chapter 3, we began with an apparently trivial problem to solve – to test whether two lines intersected. As we have seen this is not a trivial problem at all. In fact this example has illustrated a number of general points about the way that spatial data is stored and processed on the computer. What has been described is the design of an algorithm. The first point about any algorithm is that it must take into account all possible situations, including any special cases – we saw that a very simple algorithm would fail if one of our lines happened to be vertical for example. The second point is that an algorithm should be efficient i.e. it should solve the problem as quickly as possible. We may have to run a line intersection algorithm many thousands of times, and even with fast modern computers, this will take a long time if our algorithm is inefficient. There are basically two ways to make an algorithm efficient – to do whatever calculations are necessary as quickly as possible, and to do as few calculations as possible. We will see examples of the first approach in later chapters, but we have already seen that we can improve our line intersection algorithm by including simple checks to identify cases when lines could not possibly intersect, thus saving the need to run the full line intersection program on them. Thus by doing what at first sight seems to be extra work, such as finding the MER of a line, we may in fact save work overall.

In some situations, it may even be useful to store additional data in order to speed up processing. Wise (1988) described a package which allowed users to access the digitized boundaries of the wards of England and Wales. The software started by drawing a map of England and Wales on the screen, and allowing users to select their area of interest by defining a rectangular window, which would then be drawn in greater detail. Once the area of interest had been identified, the boundaries for that area could be written out as a file. This involved a lot of line intersection operations. For instance, it was necessary to decide which ward boundaries fell inside the rectangular window (and had to be drawn on screen), which fell outside (and could be ignored) and which fell across the border (and had to be cut). This meant checking every line in the database against the four lines making up the window. This was a case where the initial test of the MER made an enormous difference, and was so important that the MER for each line in the database was actually stored along with the line itself to save recalculating it each time.

3.4   CALCULATIONS ON LINES: HOW LONG IS A PIECE OF STRING?

The previous section covered an example of an algorithm for making a decision using spatial data. We also need algorithms for making calculations from spatial data and in this section we consider the simple example of calculating the length of a line. This will also illustrate the problems which can arise with computer calculations and mean that not only are the answers incorrect, they can produce incorrect decisions for algorithms such as line intersection.

Most people are probably familiar with a variety of methods for measuring the length of lines from maps. With straight line distances, a ruler can be used to measure the length on the map and a conversion is then made for the map scale. So on a 1:50 000 scale map, a line 10 cm long represents $10 \times 50\,000 = 500\,000$ cm or 5 km.

Curved lines, such as rivers, are more difficult. One method is to lay a piece of string along the line, approximating the curves and bends, and then straighten the string out and measure it with a ruler. Another is to use a pair of dividers to step along the line in short increments – the number of steps times the divider step length gives the distance.

On the computer, the method used is something of a combination of these two. We saw in Section 2.1 that our computer representation of a curved line is by a series of short straight segments as shown in Figure 3.10. These approximate the course of the line, rather like a piece of string does. The more points we use along the line, the better the approximation, but the more data we have to store. In many GIS applications, when lines are being digitized, the specification is set in terms of a maximum deviation between the straight sections and the actual line.

To calculate the length of the total line, we need to work out the length of each segment, and then add these up – this is like our divider method, but with a divider step length which varies along the line.

To calculate the length of one segment we can use Pythagoras' theorem. Many people will remember this from school as one of the basic theorems of geometry. Pythagoras proved that for any right angled triangle, the length of the three sides were always related as follows

$$a^2 = b^2 + c^2$$

(sometimes remembered via the rule – the square on the hypotenuse is equal to the sum of the square on the other two sides).

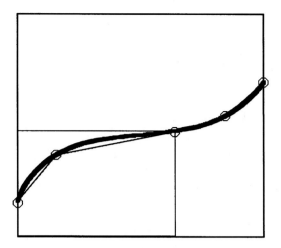

Figure 3.10 Line approximated by a series of straight segments.

In Figure 3.11, if A and B are the end points of our line segment, we can use this expression to calculate the distance between them, which is the length of the line *a* on the diagram.

Since we know the *X* coordinate of A, and the *X* coordinate of B, we can work out the length of the line b – it is simply the difference between the two values. Similarly, the length of *c* is the difference between the *Y* coordinates of A and B. We now have the values *b* and *c* and so from Pythagoras we know that b^2 plus c^2 will give us a^2 – or:

$$a = \sqrt{(b^2 + c^2)}$$

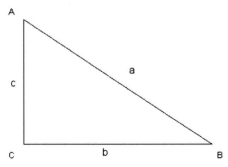

Figure 3.11 Calculating the distance between two points using Pythagoras theorem.

We can write this as it might be calculated in a computer language such as C or FORTRAN:

```
dist=sqrt((x1−x2)**2+(y1−y2)**2)
```

Here `x1`, `y1`, `x2`, `y2` are the computer variables holding the `X` and `Y` coordinates of the two points, the symbol `**2` means 'squared' (i.e. raised to power 2) and `sqrt` calculates the square root. Simple! So to calculate the length of the whole line we simply use this formula for each segment and add up the lengths. With such a simple algorithm, the issues of efficiency and the treatment of special cases which arose with the line intersection algorithm are not a problem. However, there are still situations where this algorithm will produce the wrong answer. To illustrate consider the simple example of two points with the following coordinates:

	X coordinate	Y coordinate
Point 1	0	120
Point 2	0	121

For simplicity, the X coordinates have been made the same, so we can see that the distance between the points is simply 1 – the difference between the Y coordinates. We can get our computer to check this using a simple program:

```
procedure distance(x1,y1,x2,y2)
d=square_root((x1−x2)*(x1−x2)+(y1−y2)*(y1−y2))
```

This program is very simple to code in any programming language. To illustrate the problems that can arise, a QBASIC version was run on an IBM PC. Putting in our coordinate values as above gave the answer 1 as expected. The same was true with Y values of 1200 and 1201, 12000 and 12001 etc. However, when Y values of 120000000 and 120000001 were used the answer was given as 0! Values of 12000000 and 12000001 gave 1, but the same two numbers divided by 10 (1200000 and 1200000.1) gave 0.125 instead of 0.1.

So what is happening? The problem is not with our formula, but with the way the computer stores numbers. To understand this, it is easiest to begin with our own everyday system of representing numbers. This makes do with just 10 different digits – 0 to 9. If used alone these would only be capable of representing 10 numbers, which would be of little use except for counting very small flocks of sheep (This is not as silly as it sounds – there are primitive tribes who manage with three numbers – 1, 2 and 'many'). However, we use our 9 digits to represent all the numbers we need by combining them together. Thus in a number like 245, the 2 means '2 lots of a hundred', and the 4 means '4 lots of ten'. We extend the range of numbers we can deal with by using a decimal point to indicate fractions of a whole number – hence in 24.5 the 5 means '5 lots of 0.1'.

Another way of representing this is to see each number as a sum:

Digit		Position		
2	×	100	=	200
4	×	10	=	40
5	×	1	=	5
				245

Moving one position to the left increases the importance of the digit by a power of ten. The binary system used by computers is exactly the same, except that there are only two digits, and each position increases its value by a power of 2. Hence, the number 110 in binary is equivalent to 6 in decimal:

Digit		Position		
1	×	4	=	4
1	×	2	=	2
0	×	1	=	0
				6

In order to represent large numbers the computer must use groups of digits. A single digit in the binary system is called a bit (short for Binary digIT). A group of 8 digits is called a byte. A byte can be used to store whole numbers (without a decimal point) between 0 and 255. Larger groups of bits are usually called words, although they can actually vary in size from 16 to 64 (and on some computers 128) bits. The largest whole number which can be stored in a word can be calculated using the formula

$$2^n - 1$$

where n is the number of bits. The most usual word length is 32 bits, which can store a number up to 4 294 967 294 (just over 4 billion). If we want to be able to store negative numbers too, then one of the bits in the word is reserved for this purpose. This is normally the leftmost bit, and if this is 0 the number is positive, if it is 1 the number is negative.

This is fine for whole numbers (or integers). But what about numbers with a decimal point in (sometimes called real numbers)? Let us return to the world of base 10 numbers again for an explanation.

With a number like 24.5 we have to store the digits 2, 4 and 5 and also the fact that there is a decimal point after the first two digits. In fact, rather than store the position of the decimal point, we move it so that it is to the left of the first digit, and then record how many places it has been moved. So 24.5 is actually stored as:

$$24.5 = 0.245 \times 10^2$$

This system is widely used in the scientific literature because it is a convenient way of dealing with very large or very small numbers. For example, the mass of an electron at rest can be represented as $0.9109534 \times 10^{-30}$ g which is far easier to deal with than a string of 30 zeroes followed by the digits 9109534! On the computer, this method of representation is called floating point, because the decimal point 'floats' to the front of the number and the term floating point is often used to refer to numbers which are not integers.

To store a number in floating point we actually have to store two numbers – the actual digits themselves (called the mantissa) and the number of 10s they must be multiplied by (called the exponent). In the world of pen and paper, there is no limit to the length of either the mantissa or the exponent. However in the world of the computer we are limited to what we can store in our word. This is like having 32 boxes, each of which can hold one digit. Some of the boxes will be used for the mantissa and some for the exponent. If we split them evenly, we can store numbers with 10 000 000 000 000 000 digits in them (the maximum exponent size) but in the mantissa there is only room to store the first 16 of them. It is more usual therefore to reserve more spaces for the mantissa than the exponent. Even with a 2/30 split we can store numbers with 100 digits, but only record the first 30.

Exactly the same principle applies in the binary system, but in this case each box can only hold one bit rather than a decimal digit. The same problem also arises – this technique allows us to store very large or very small numbers, but only the first few digits of those numbers can be recorded – the rest are simply lost. So when we attempt to calculate distances using coordinates like 1200 and 1201, we are safe, but as soon as the numbers get up to the range of 12000000 and 12000001 the results become unreliable.

These problems are compounded whenever calculations are performed – the loss of digits in the numbers means the answer is slightly wrong, and if this is then used in another calculation that answer will be even more wrong and so on. It is not just calculations such as distance which are affected. The line intersection algorithm described in Section 3.2 was based on a series of calculations, any of which could be affected by computer precision problems. For example, a key part of the code for line intersection is the final set of tests to see whether the computed intersection point lies on both the line segments:

```
if(xs1-xp)*(xp - xe1)>=0 and
  (xs2-xp)*(xp - xe2)>=0 and
  (ys1-yp)*(yp - ye1)>=0 and
  (ys2-yp)*(yp - ye2)>=0 then
    print 'Lines cross at' xp,yp
else
    print 'Lines do not cross'
```

Let us consider the first line, which tests whether the X coordinate of the intersection point (xp) lies between the X coordinates of the two ends of the first line segment (xs1, xe1). xp has been calculated earlier in the algorithm. If the coordinate values of the line segments contain more than 7 significant digits, and their values are very similar, then xp may appear to be the same as either xs1 or xe1, when in fact it is not. This will make the result of the overall calculation 0, making it appear that the intersection point falls on this line segment, when it may actually fall just outside it. Conversely, xp may actually be exactly the same as either xs1 or xe1, but because of the lack of precision in the calculations, appears to be slightly different. Again, this could lead to the wrong conlcusion being reached in the test.

These problems can easily arise in real world applications. For example, in Great Britain the National Grid is commonly used for coordinates. This has its origin somewhere near the Scilly Isles, so that by the time you reach Scotland, the Y coordinates of points will be around 800 000 – in the Shetlands they exceed 1 000 000 metres, which is on the limit of standard computer precision.

One solution is to use a local origin for the coordinates of points. In the example of the Shetland Islands, all Y coordinates could have 1 000 000 subtracted from them, so that the important variations between positions within the area fall within the computer precision. An alternative solution is to use what is called double precision, if this is supported by the GIS package. This uses a longer word for each number, increasing the number of significant digits which can be stored, but also doubling the amount of storage space required. GIS designers can help avoid these problems by designing algorithms which explicitly deal with issues of precision. This may involve replacing all the simple tests for equality (==) and inequality (>, > = etc.) with more complex tests which include tolerance values rather than testing for absolute equality or inequality.

FURTHER READING

The discussion of the line intersection problem in this chapter is largely based on the material from Unit 32 of the NCGIA Core Curriculum, and on the papers by Douglas (1974) and Saalfeld (1987). De Berg *et al.* (1997) have a good discussion of algorithms for the multiple line intersection problem in Chapter 2, and of windowing in Chapter 10. Nievergelt (1997) provides an excellent introduction to the development of computational geometry, and uses the example of finding the nearest neighbour among a set of points to illustrate some of the features of designing algorithms for handling spatial data. The problems caused by limited computer precision are well illustrated by the example discussed by Blakemore (1984) in the context of locating points in polygons. Greene and Yao (1986) have proposed a solution to these problems, by explicitly recognizing that vector data has a finite resolution and that operations such as line intersection must take this into account. Worboys (1995) provides a brief summary of their approach. Cormen *et al.* (1990) feel that this problem is so serious, that the line intersection algorithm should be handled completely differently. In their approach, the endpoints of the line segments are compared to identify whether the lines intersect, and the calculation of where the intersection point falls comes later. This means that the decision about intersection is based on comparisons of the original coordinates, not on the result of calculations. Schirra (1997) provides a much more extensive discussion of this issue, pointing out some of the weaknesses of standard solutions such as including tolerance values in tests for equality.

The paper by Wise (1988), while mostly about a particular piece of computer hardware, also includes a discussion of the use of the MER in designing a system which allowed users to select windows from a national data set of area boundaries.

4 Vector algorithms for areas

4.1 CALCULATIONS ON AREAS: SINGLE POLYGONS

In Section 2.2, we have seen that area features are stored in GIS as polygons, represented using a single closed line, or a set of connected links. In this section, we are going to consider some of the fundamental operations involving areas in GIS, starting with finding out how large an area is. For the moment, we will assume that the polygon is stored in the simplest possible form – as a line which closes on itself like the example in Figure 4.1.

This line is stored as a series of 5 points which form the ends of the five straight segments. Three of these form the upper part of the boundary, and two the lower half. If we calculate the area under the 'upper' segments, and subtract the area under the 'lower' segments we will be left with the area of the polygon itself, as shown diagramatically in Figure 4.2.

This is fairly easy for us to understand visually, but we must remember that the computer does not 'see' the area in the way we do – all it has is a series of X, Y coordinates for each of the 5 points on the polygon boundary. I will describe how we can determine which parts of the boundary are above and which below the figure in a moment – first we need to know how to calculate the area under a line segment, so it is back to the maths classroom once more!

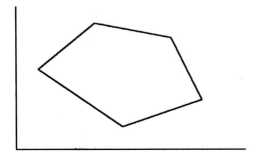

Figure 4.1 Single polygon.

Figure 4.2 Calculating the area of a polygon – area beneath 'lower' segments is subtracted from area under 'upper' segments.

Let us start with the first line segment, shown in Figure 4.3. The area under this line can be divided into two parts, as shown by the grey lines. The lower part is a rectangle, with width $(x_2 - x_1)$ and height y_1, so its area is simply:

$$(x_2 - x_1) \cdot y_1$$

The upper part is a triangle, whose area is half the area of the small rectangle as shown. The width of this rectangle is $(x_2 - x_1)$ as with the first one. Its height is $(y_2 - y_1)$ and so half its area will be:

$$(x_2 - x_1) \cdot \frac{(y_2 - y_1)}{2}$$

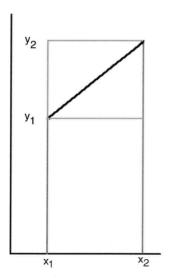

Figure 4.3 Calculating the area under a single line segment.

If we add these two areas together we get the following formula:

$$(x_2 - x_1) \cdot \left[y_1 + \frac{(y_2 - y_1)}{2} \right]$$

which can be simplified to the following:

$$(x_2 - x_1) \cdot \frac{(y_1 + y_2)}{2}$$

We can apply the same formula to each line segment replacing (x_1, y_1) and (x_2, y_2) with the appropriate values for the start and end points of the segment. So how do we know which segments are below the polygon, and need to have their area subtracted from the total? To answer this, let us simply apply our formula to the segment between points 4 and 5 and see what happens (Figure 4.4). In this case, the formula for the area will be:

$$(x_5 - x_4) \cdot \frac{(y_5 + y_4)}{2}$$

The first term will give a negative value, because x_5 is smaller than x_4 and so the area under this line segment will have a negative value. In other words, by simply applying the same formula to all the line segments, those under the polygon will have a negative area while those above will have a positive value. This means that we do not have to take any special action to identify which segments are above and which below the polygon – we simply apply

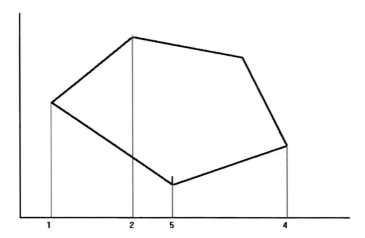

Figure 4.4 Area under line segments which are 'above' and 'below' the polygon.

our formula to each segment in turn and by summing them as we go we will arrive at the area of the polygon.

This simple algorithm will work as long as the polygon boundary is stored as a sequence of points in a clockwise direction – if the points are stored in an anticlockwise direction, the answer will be numerically correct, but be negative rather than positive!

Of course not all polygons have such simple boundaries as this one. However, as long as the boundary is stored as a clockwise sequence of points, the algorithm will still work correctly no matter how convoluted or strangely shaped the boundary.

4.2   CALCULATIONS ON AREAS: MULTIPLE POLYGONS

In many GIS systems, area features are not stored using simple polygons, but using the link and node structure described in Section 2.3. This data structure is particularly useful when a GIS layer contains a series of polygons which border each other, because it is only necessary to store the shared boundaries between polygons once, as shown in Figure 4.5.

This shows two polygons, A and B, whose boundaries are made up of links (numbered 1, 2 and 3) which meet at nodes (indicated with circles). Each link is stored as a series of points (for which we have an X and Y coordinate) which define a series of straight line segments. In addition, we also know which node the line starts and ends at, and which polygon is on its left and right.

As we saw in the previous section, if the boundary of a polygon is a single line running in a clockwise direction, the area of the polygon can be

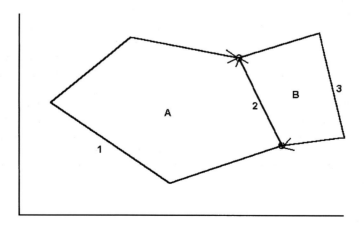

Figure 4.5 Storage of two neighbouring polygons using the link and node data structure. The arrows show the direction of digitizing of each link.

calculated by working out the area under each line segment using the following formula:

$$(x_2 - x_1) \cdot \frac{(y_1 + y_2)}{2}$$

Here x_1, y_1 are the coordinates of the start of the line and x_2, y_2 the coordinates of the end of the line. These areas will be positive for lines 'above' the polygon and negative for lines 'below' the polygon, so by summing them the area of the polygon itself is calculated.

In Figure 4.5, instead of a single line we have a set of links forming the boundary. However, this does not really affect our formula, which considers one line segment at a time – as long as we can connect our links into a clockwise boundary, the method will work as before. If we look at polygon B, we will see that its boundary is made up of links 2 and 3 which do indeed go clockwise around the polygon. In this case, we can simply apply the algorithm to the two and we will get the area of polygon B.

However, this is not true for polygon A because link 2 is going in the wrong direction. We can see this from the diagram, but of course all the computer has to go on is the information we have stored about the links. We thus have two problems to solve – how can the computer tell whether a link is going the right way around a polygon, and if it is not, what can be done about it?

The answer to the first lies in the topological data stored for each link. If a line circles an area in a clockwise direction, then the area will be on the right hand side of the line – the same logic also applies to the links in a link and node structure. In fact, you may remember we used the same rule when checking the accuracy of street data in the DIME system. Applying this rule, we will be able to detect that link 2 is the wrong way round for calculating the area of polygon A since A is on the left of this link. For calculating the area of B the link is in the right direction, of course.

One option for dealing with this is to reverse the direction of link 2 when doing the calculations for polygon A. However, if we use this approach we will apply our formula to link 2 twice – once while calculating the area of polygon A, and once while calculating the area of polygon B. The only difference is that link 2 is 'above' polygon A but 'below' polygon B – in the first case the area under the link must be added to our sum, in the second case subtracted. A better approach is to calculate the area under the link once, and then use the topological information to decide whether to add this or subtract it from the area of each polygon.

Let us assume the coordinates of the line as shown in Figure 4.5 are as follows:

	X	Y
Start	4	1
End	3	2

When we enter these figures into the formula:

$$(3 - 4) \cdot \frac{(1 + 2)}{2}$$

we get the answer -1.5. The result is negative, because the start point is to the right of the end point and so has a larger X coordinate. Link 2 is pointing the right way to form part of a clockwise boundary around polygon B, and the computer knows this because B is on the right of link 2. The same logic applies in this case as with the simple polygon described in Section 4.1 – the negative area ensures that the area under link 2 will be subtracted from the running total. When we try and calculate the area of polygon A, we need to know the area under link 2. This will still be 1.5, but now it must be added to the sum, and not subtracted. Since we know it is running anticlockwise around A (because A is on its left) all we need to do is take the result of the formula and reverse the sign.

This example has used just one link. The general rule is as follows. Take each link in turn and apply the formula to calculate the area under that link. For the polygon on the right, the link is running clockwise and so the area can be used exactly as calculated (whether this is negative or positive). For the polygon on the left, the link is running anticlockwise, so the sign of the area is changed (from negative to positive or from positive to negative) and then the answer added to the sum for that polygon. If you want to check that this works, work out what would have happened if link 2 had been digitized in the other direction – the area should still be negative when link 2 is considered as part of polygon B, but positive when it is part of polygon A.

Some links will only border one 'real' polygon – on the other side of the line will be the 'outside' polygon. An interesting side effect of this algorithm is that if the same rule is followed for this outside polygon as for all the others, it receives an area which is the total of the areas of all the other polygons, but with a negative value.

4.3   POINT IN POLYGON: SIMPLE ALGORITHM

Now we will turn to another of those problems which seems simple but which is difficult to program a computer to solve – determining whether a point lies inside or outside a polygon. A typical example of the use of this algorithm is the matching of address-based data (assuming we can determine a location for the address) with area data such as data from a population census. Typical examples will try and match several points to several polygons, but for simplicity we will begin with one point and one polygon, as shown in Figure 4.6.

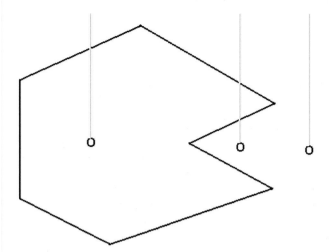

Figure 4.6 Point in polygon test.

As with the line intersection problem we considered earlier in this series, it is very simple for us to see that only the leftmost point lies within the polygon. However, all the computer can 'see' is the position of the points (as X, Y coordinates) and the location of the points defining the polygon boundary. A number of methods have been developed to test if a given point lies inside a polygon. In the one which is most commonly used, an imaginary line is constructed extending out in any direction from the point – if this line intersects the boundary of the polygon an odd number of times, the point is inside the polygon. In Figure 4.6, the line from the first point intersects the boundary once and so this point is inside the polygon. The line from the second point intersects twice and so this point is outside. The line from the third point does not intersect at all. Mathematically speaking, zero can be considered an even number because there is no remainder when it is divided by 2, and so the test works in this case too. In simple terms the algorithm is:

```
1  n=0
2  for each line segment
3    test intersection with 'ray' from point
4    if lines cross, n=n+1
5  if n is odd, point is in polygon
```

Each line segment in turn is tested for intersection with the 'ray' from the point and a variable called n is used to count these intersections. An interesting detail of this algorithm is the method used to determine whether the number of intersections indicate that the point is inside or outside. The method above counts the intersections but then must determine whether the total is an odd or even number. It is also possible to write the algorithm to give the answer directly by using what is sometimes called a 'toggle' – a variable which switches between two possible values:

```
1  n=-1
2  for each line segment
3    test intersection with 'ray' from point
4    if lines cross, n=n*(-1)
5  if n=1, point is in polygon
```

In this example variable n is set to -1. If an intersection is found n is multiplied by -1 changing its value to $+1$. If a second intersection is found n is again multiplied by -1 changing its value back to -1. The value of n therefore toggles between the values of -1 and $+1$ each time an intersection is found – an odd number of intersections will leave the value at $+1$.

This sort of detail will not affect the speed of the point in polygon test very much, but the way in which the line intersection test is done will. In Chapter 3, we have seen how to test whether two lines intersect and so it would seem simple enough to use this program to test each segment of the boundary line in turn against our ray. This would certainly work, but as with the line intersection test itself, there are a number of things we can do to produce a much faster method.

The first of these is to use the MER test. As shown in Figure 4.7, we can begin by testing the points against the smallest rectangle which

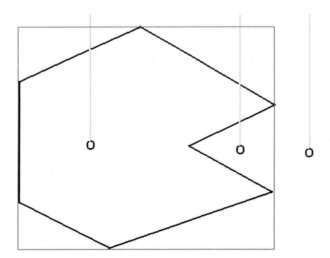

Figure 4.7 Use of the MER for the point in polygon test.

encloses the polygon – a point which falls outside this rectangle cannot lie inside the polygon and there is no need to perform any further tests. As this example shows, it is possible for a point to lie inside the MER, but outside the polygon itself. However, if we are testing a single point against a large number of polygons (as with the example of finding which census area an address lies in) this initial test will help us reject a large number of the polygons very quickly. Once we have our candidate polygons, there are other things we can do to improve the speed of the line intersection test.

The standard line intersection method first determines the equations defining the two lines, calculates the point at which infinite lines defined by these equations intersect and then tests whether this point lies on both of the actual, finite lines being tested. However in this case, we can determine the characteristics of one of the lines in question – the ray extending from the point – and as Figure 4.6 shows, a vertical line is used. This alone would speed up the standard line intersection test, since we know the X value of the intersection point and would only have to work out the Y value.

We can make further improvements by developing a special algorithm for line intersection in this particular case. Since the ray is vertical, a line segment can only intersect with it if the ends of the segment lie on either side of the ray. If we denote the X value of the ray as xp (since it is the X coordinate of our point), intersection is only possible if xp lies between x_1 and x_2 and one way to test for this is as follows:

$$((x_1 < x_p) \quad \text{and} \quad (x_2 > x_p)) \quad \text{or} \quad ((x_1 > xp) \quad \text{and} \quad (x_2 < xp))$$

Figure 4.8 shows what happens when we apply this test to the leftmost point in Figure 4.6. Two line segments have passed the test, because both intersect the infinite vertical line which passes through the starting point of the ray. We are only interested in those segments which intersect the finite vertical line which passes upwards from the starting point. In order to find out which these are we will need to perform the full line intersection calculations, and then test whether the intersection point is above or below the starting point of the ray. The advantage of our initial test is that we will only have to do this full set of calculations for a small proportion of the segments making up the polygon boundary.

In making this test, we have to be careful about the special case where the ray passes exactly through one end of the line, as shown in Figure 4.9. In this case, Xp is equal XB and if we perform the test as shown above, neither line will pass. The test must therefore be modified to become

$$(x_1 = xp) \quad \text{or} \quad ((x_1 < xp) \quad \text{and} \quad (x_2 > xp)) \quad \text{or} \quad ((x_1 > xp) \quad \text{and} \quad (x_2 < xp))$$

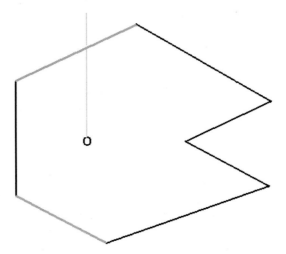

Figure 4.8 Line segments selected by initial test on X coordinates.

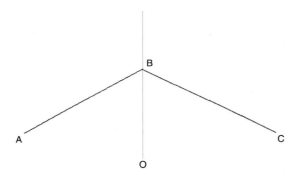

Figure 4.9 Special case of a point lying directly beneath a vertex on the polygon boundary.

Note that we only test one end of each segment to see if it is the same as Xp. In the example above, the first segment will fail the above test because X_1 in this case will be the value for XA – however for the second segment X_1 will be equal to XB and so the segment will pass the test. If we tested both ends of the segment for equality with Xp, both segments would pass the test and we would still miscount.

In fact, the test as described above will run into problems because of the limited precision of floating point operations, as described in Section 3.4. In particular, testing for exact equality $(x_1 = xp)$ will be particularly sensitive, and it is far safer to test for x_1 being within a very small distance of xp. To do this we replace the term in the first brackets with:

$$(x_1 > (xp - d) \quad \text{and} \quad x_1 < (xp + d))$$

where d is a very small distance.

These alterations to the basic original algorithm will not make much difference if we have one point and one polygon, but in more typical uses of the algorithm there will be several points and several polygons. For example, in a study of disease incidence in Sheffield, staff at Sheffield University used a GIS to find which of the 1100 census EDs each of 300 patients lived in. This meant testing each of 300 points against 1100 polygons – on average a match will be found for each point halfway through the set of polygons, but this still gives 165 000 tests to be performed. However, if we apply our simple MER test then in most cases only 3 or 4 polygons will be possible candidates meaning we only have to run the full test about 1200 times. The saving in time will more than compensate for the extra work of doing the MER tests.

4.4 ... AND BACK TO TOPOLOGY FOR A BETTER ALGORITHM

So far, we have only considered how to decide whether a single point lies inside a given polygon or not. However, rather than a single polygon, it is more usual to have a GIS layer containing numerous polygons, stored using the link and node method described in Section 2.3. What is needed is to find which of the many polygons a point falls into.

One approach would be to test the point against each polygon in turn. We would have to construct the boundary of the polygon, by retrieving all the relevant links, find the MER and then if the point fell inside this, carry out the full point in polygon test.

The problem with this approach is that most links would be processed twice. The example in Figure 4.10 shows a point in polygon 8. The link between polygons 7 and 8 starts at node A, runs through points b, c, d and e and ends at node F. This link would first be tested as part of the boundary of polygon 7. The test point just falls inside the MER of polygon 7, and so the full point in polygon test would be performed – each of the segments of this link (A to b, b to c etc.) would be tested in turn against the ray and in two cases intersections would be found – an even number of intersections indicate a point outside the polygon, and so the algorithm would move on to polygon 8.

The point clearly falls in the MER of polygon 8, and so exactly the same set of operations will be carried out on the A–F link, and the same two intersections found. (Further intersections will be found in this case, with other links forming the border of polygon 8.) So is there a way of avoiding this double processing? There is and again the key is in the use

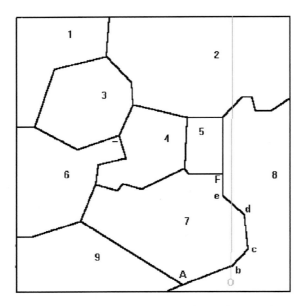

Figure 4.10 Point in polygon test for multiple polygons. The point is indicated by the grey circle and test ray by the vertical grey line.

of topology. We have already seen that polygon boundaries can be stored using links which join at nodes, an idea developed by applying Poincarés model of maps as topological graphs. However, this model also means that whenever lines cross, they MUST have a node at the crossing point. To explain what this means, and why it is important, consider Figure 4.11.

The upper diagram contains two polygons which overlap but which do not have nodes where their boundaries cross. The problem with this is that the rules which we used for checking the accuracy of a digitized map no longer apply. For example, in checking the DIME file, it could be assumed that each line formed part of the border between two areas, but in the upper diagram the same line can sometimes form the border between polygon 2 and the inside of polygon 1, and also between polygon 2 and the 'outside world'. In order for the simple topological rules to work, GIS systems use what is called 'planar enforcement' – this means that lines are all considered to lie in the same plane, and so if they cross, they must intersect – one line can't go over another one.

This is illustrated in the lower diagram, where nodes have been placed where the lines cross. This actually creates TWO new polygons and modifies the two original ones. Polygons 1 and 2 in the upper figure are each split

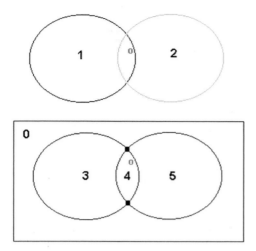

Figure 4.11 The consequence of the planar enforcement rule.

into two parts, and the area in common between them becomes a polygon in its own right (polygon 4 in the diagram). In addition, since the rules state that each link **must** lie between two polygons, there has to be an 'outside' polygon which is labelled 0 on the map.

The difference between the two is very relevant to the point in polygon problem. Consider the point indicated by a grey circle – in the upper diagram this lies in both polygon 1 and polygon 2 – in the lower diagram it is only in polygon 4. In fact, if we enforce the rule of planar enforcement, then the rules of topology tell us that (a) any point MUST be in a polygon (even if this is the 'outside' polygon); and (b) that it can only be in one polygon. (Strictly speaking, a point can also lie on the border between two polygons, but this is a detail which can be safely ignored for the purposes of explanation – for the design of a real algorithm, it would have to be dealt with in a rigorous manner).

Given these facts, if we project a ray upwards from a point, then the first line which it crosses must form the border of the polygon which contains the point. This means there are now two stages to our point in polygon test:

1 Find the first link which the ray intersects.
2 This link will lie between two polygons, so we will need to discover which of these contains our point.

As with the original version of the test, the key is the fact that the ray is vertical – this means that if we find all the links which the ray crosses, the first one is simply the lowest. The simplest approach to finding which links intersect the ray is to test them all, but we can speed up this process if we calculate the MER for each link, and use this to filter out the candidate links first. Once we have found the lowest link, we know that the polygon containing the point is the one below this link – but how do we find out which polygon is below and which above? Once more topology comes to the rescue as the example in Figure 4.12 illustrates.

In this example, the ray intersects the same link twice but by looking at the Y values of these intersections, it is easy to decide which is lower. We know the overall direction of the link because we have stored the 'from' and 'to' nodes – the direction is shown with an arrow on the diagram. This means we can tell whether the part of the link where the ray intersects runs from east to west or west to east – in this case it is EW, since the X value of the start of the segment is greater than the X value of the end. With an EW segment, the polygon on the left (C in this case) will be below the segment, and thus will be the polygon we want.

Note that in this case we do not have to count intersections at all. Our topological rules of planar enforcement tell us we have a set of polygons covering the area (including the 'outside' polygon) and so the point must lie in one of them. The problem is simply to determine which one.

Hopefully, this discussion of vector data structures and algorithms will have given you a good understanding of some of the issues which arise in writing a piece of GIS software. We have seen that algorithms for apparently simple tasks can be very complicated. We have also seen that the data structure which is used also has a key role to play in algorithm design, and this is an issue which will arise repeatedly in the rest of the book. Throughout the chapters on vector algorithms there has been a concern with the speed of execution of algorithms, and we have seen some of the strategies which have been used to make algorithms faster. So far we have not made any formal assessment of how much improvement we can expect from these strategies, and so we will turn to this in Chapter 5.

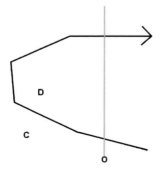

Figure 4.12 Testing which polygon is below a link.

FURTHER READING

The calculations for polygon areas are covered in Burrough and McDonnell (1998). A brief account of point in polygon is contained in the NCGIA Core Curriculum unit 33, while Unit 34 covers the polygon overlay operation Both De Berg *et al.* (1997) and Worboys (1995) give details of other algorithms for the point in polygon problem, and describe algorithms for polygon overlay. Huang and Shih (1997) compare a number of different points in polygon algorithms concluding that the characteristics of the polygons themselves have a strong effect on how quickly the algorithms run. White (1978) gives some background to the WHIRLPOOL algorithm, which was the first efficient implementation of polygon overlay in a vector GIS. Teng (1986) discusses the importance of both topology and the preprocessing of data in handling the polygon overlay operation.

5 The efficiency of algorithms

5.1 HOW IS ALGORITHM EFFICIENCY MEASURED?

We have seen that an important part of designing GIS algorithms is to make them run as quickly as possible. In the case of the simple point in polygon test for example, testing the point against the MER of the polygon could save a lot of unnecessary line intersection tests. A lot of things can affect how quickly an algorithm will run such as the speed of the computer and how many other tasks are running on it at the same time. However, many of these will be difficult to measure, so in assessing how efficient an algorithm is, computer scientists concentrate on what is called algorithm complexity – how the performance of the algorithm is affected by the amount of data to be processed. For example, if we have twice as much data to process how much longer will the algorithm take to run? Twice as long? Four times as long?

To see how algorithm complexity is assessed, it is easiest to consider an example of an everyday task – looking up a name in the phone book. The simplest method would be to start at the beginning of the book and work our way through until we found the name we want. If we are lucky, we may find the name quickly – if we are unlucky we may go through the whole book. Let us call this the 'brute force' algorithm. If there are 64 entries in the book, then on average we will search half of them before finding the correct entry – this means making 32 comparisons between the entry and the name we are looking for. If the size of the book doubles, the average number of searches also doubles. The complexity of this algorithm is therefore directly related to the size of the problem. There is a standard notation used by computer scientists to express this called 'big O notation'. In this case, we would say that the brute force algorithm has $O(n)$ computational complexity – the complexity is directly related to the size of the problem, denoted conventionally by the letter n. The capital O stands for Order and we could equally well say that the algorithm has order n computational complexity. Note that $O(n)$ does not mean it will always take n operations to solve a problem – even with the brute force algorithm we may get lucky and find the answer straight away. It is used to indicate the complexity of the problem under the worst case.

The brute force algorithm is a poor one, because it does not take advantage of the fact that the names in the phone book are in alphabetical order. Assuming the name we were seeking was 'Wise', most people would start by turning to the back of the book since W is at the end of the alphabet. Once they found the W section, they would look in the first third, since 'I' comes about a third of the way through the alphabet and so on – in effect, rather than search the whole book, the best approach is to successively narrow down the search to the right part of the book.

To program a computer to perform the same sort of task, we would use what is called a binary search. Rather than start with the first entry in the list, we begin with the middle one – with a list of 64 this would be number 31 or 32 (with an even number of entries there is no middle one but we would devise a simple rule to pick one of the two entries which straddle the middle of the list). We compare the name with 'Wise'. If they match, the job is done. If 'Wise' comes later in the alphabet, we know we need to look in the second half of the list and vice versa. We therefore repeat this process with the appropriate half of the list. At each stage of the search we split the list in half, and identify which half the item we are looking for is in. We start with 64 entries, which are split into 2 groups of 32. One of these is split into 16, and then 8, 4, 2 and 1 – by the time we have only one entry we will either have found the name we are looking for, or will know that it is not in the list. This means that the maximum number of comparisons we will have to make is just 6. If the size of the list doubles, the maximum number of comparisons only rises to 7 – one more subdivision of the list into two halves.

The number of comparisons for a binary search is related to the number of times we have to divide the list into two halves before we are left with a single entry. If the number of items is n, this is the same as the number of times we have to multiply 2 by itself in order to get n. In the first example n was 64, which is $2 \times 2 \times 2 \times 2 \times 2 \times 2$ or 2^6 (two to the power of six). We had to divide the list into 2 six times at most to find our entry. If n is doubled to 128, this means the number of searches goes to 7, and 128 is 2^7. The complexity of this algorithm is therefore related to the number of times we have to multiply 2 by itself to get n. Fortunately, there is a simple mathematical term for this – the logarithm to the base 2 of n. In mathematical notation,

If $n = 2^m$, $\quad m = \log_2(n)$

So the complexity of the binary search algorithm is $O(\log_2(n))$. Logarithms can be calculated using any base – logarithms to base 10 used to be widely used to perform complex calculations in the days before the pocket calculator. However, in terms of assessing the efficiency of different algorithms, the base used for the logarithms is not important (for reasons outlined in Section 3 of this chapter) so the efficiency of the binary search is reported as $O(\log n)$.

The binary search algorithm will only work if the list is in order. This is in alphabetical order for our telephone book example, but if we were searching other items, we might need the items in numerical order or date order. So to make it a general purpose algorithm we should really include the first step of checking whether the list is ordered and sorting it if not. Checking is an $O(n)$ operation. We start at the first item and see whether it is smaller or larger than the second – let us assume it is smaller. Then we look at the second item and check whether this is smaller than the third and so on down the list. If we find any item out of sequence the list is unordered, and to discover this we look at every item except the last (which does not have anything below it in the list to be compared with).

To see how we might sort the list, first of all imagine a related problem. Assume we have an ordered list, and we want to add one more item to it. To find out where it should go we perform a binary search. If the item is already in the list, this will return its position. If it is not in the list we will find the position where it ought to go. In order to sort a complete list we use the same approach. We begin with our unordered collection of items, and we set up an empty list. We then add the items to the list one at a time. In the first case, the first item is the only thing in the list so the operation is trivial. The second item can only go before or after the first, and this is also simple. As the list grows however, the binary search will be the fastest way of finding the correct place in the sequence for the next item. We know a binary search runs in $O(\log n)$ time, and since we will have to run one for each item to be added, the total complexity of this algorithm is $O(n \log n)$.

To find the efficiency of the total algorithm then, we simply add together the three terms:

$O(n)$ for checking whether the list is sorted
$O(n \log n)$ for sorting it
$O(\log n)$ for performing a search
Hence $O(n + n \log n + \log n)$

In the function within the brackets, $n \log n$ is the dominant term, and so the complexity of the overall algorithm is reported as $O(n \log n)$. It may seem as if we are ignoring important details in only reporting the dominant element in the function. However, the difference between an algorithm which performs $n \log n$ operations and one which performs $n \log n + \log n$ operations is much less important than the difference in performance between an $O(n \log n)$ algorithm and one which is $O(n^2)$ as shown in Figure 5.1. This shows how the value of some of the common functions changes as n changes, and also the terms used for some of the functions.

n	Linear $O(n)$	Quadratic $O(n^2)$	Logarithmic $O(\log n)$	$O(n \log n)$
1	1	1	0	0
10	10	100	1	10
100	100	10000	2	200
1000	1000	1000000	3	3000
10000	10000	100000000	4	40000

Figure 5.1 Value of some of the functions commonly used in reporting algorithm complexity.

Big O notation is not only used to assess the processing efficiency of algorithms – it can be applied to any aspect of algorithms. When we look at raster GIS in Chapter 6, we will see that an important aspect of different ways of storing and processing raster data is the trade off between the amount of memory needed and the speed of answering queries, and both these can be assessed using big O notation.

5.2   THE EFFICIENCY OF THE LINE INTERSECTION ALGORITHM

Let us look at the application of big O notation by assessing the efficiency of the line intersection algorithm. Figure 5.2 shows two lines, each represented by a series of line segments – 7 in the case of the thin line, and 3

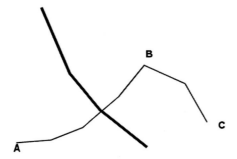

Figure 5.2 Line intersection example.

in the case of the thick line. In Section 3.2, we saw what was necessary to determine whether two line segments intersected. Many GIS operations require the system to compare two sets of lines and determine which of them intersect and where – note that since lines can curve, two lines may actually intersect more than once along their length. The brute force approach to this problem is to test every line segment against every other line segment – if there are n segments, we will perform n^2 intersection tests and this algorithm is of $O(n^2)$ computational complexity.

If we use a MER to identify candidates for intersection, then the number of intersection tests will be greatly reduced. It is difficult to say exactly how many intersection tests we will have to perform because it depends on the configuration of lines as shown in Figure 5.3. At one extreme we might have a set of lines in which all the MERs overlap, and we still have to perform n^2 intersection tests. At the other extreme, we may have a set of lines in which none of the MERs intersect, and we have to perform no intersection tests at all. These worst and best case scenarios are extremely unlikely in practice of course. A more typical scenario is shown on the right of Figure 5.3, in which each line is near to a small number of other lines.

In the typical scenario there are two ways in which the number of potential intersection tests could increase. We could be dealing with a larger study area, with lines at approximately the same density – in this case the number of tests would simply increase as n increased. Alternatively, we could be dealing with a similar sized area, but containing more lines. Each line would therefore have more neighbours and if the number of extra neighbours was roughly related to the number of extra lines, we might have $2n$ tests to perform.

However, out overall line intersection algorithm is still $O(n^2)$ because although the MER test reduces the number of full line intersection tests we perform, we still have to compare every MER with every other MER, and this is n^2 comparisons. If we could reduce the number of comparisons then our algorithm would be improved. The key to good algorithms is often to

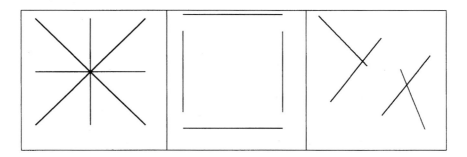

Figure 5.3 Line intersection using MER (from left to right) worst case, best case, average case.

determine how to avoid doing unnecessary work. In the case of searching for names in a list, we saw that the key was to first sort the names into order because then we could employ an efficient method of searching the list. The analogy to the line intersection problem is that we need some way of identifying which MERs are near to each other, and need to be tested for intersection, and which are not and therefore do not need to be tested. This idea of identifying spatial neighbours is a key one in the design of spatial data structures and algorithms, and we will return to it in Chapter 8. In the case of the line intersection problem, one approach is to sort the MERs into order based on their maximum Y coordinate values. The algorithm then proceeds by sweeping an imaginary horizontal line down the map from top to bottom. Whenever it passes the upper part of an MER, this line is added to the current set, and tested against all the other lines in the set. When the sweep line passes the lower coordinate of an MER, the line is removed from the set and is not used in any further tests. This strategy is commonly used in computer graphics and GIS and is called a plane sweep algorithm. A full algorithm of this sort is described by De Berg *et al.* (1997) who show that it has $O(n \log n + k \log n)$ complexity, where k is the number of intersections found.

5.3 MORE ON ALGORITHM EFFICIENCY

We have seen how big-O notation can be used to identify weaknesses in algorithms and distinguish between efficient and inefficient ones. We will end this chapter by providing a little more detail on what this notation means and why we are able to ignore some terms in reporting efficiency, and introducing some related measures of efficiency.

In Section 5.2, we saw that when we say that an algorithm has $O(n^2)$ efficiency, this means that the relationship between the number of operations the algorithm performs and the problem size is approximately quadratic. To understand what this means, let us take the example of the search algorithm from Section 5.1. We saw that the number of operations needed to find an item in a potentially unsorted list was $n + n \log n + \log n$. However, we reported that efficiency of the algorithm as $O(n \log n)$. If we tabulate the values of these two functions for a few values of n, we will see that $n \log n$ is always smaller than $n + n \log n + \log n$ (Figure 5.4). So it seems as if we are exaggerating the efficiency of our algorithm.

However, we are really most interested in large values of n. Consider our searching problem again. If we have to find a name in a list of 5 names, it does not much matter how we go about it. The problem is so small, that even an inefficient method will be quick enough. If we have 5000 names, the situation is different however. It is worth the extra effort of sorting the names into order, because this will greatly speed up the search process. Sorting 5 names into order would take almost as long as simply going through the list from 1 to 5 looking for the name we want, and actually makes the process slower rather than quicker.

n	$\log n$	$n \log n$	$n + n \log n + \log n$	$2(n \log n)$	$0.5(n \log n)$
2	1	2	5	4	1
4	2	8	14	16	4
8	3	24	35	48	12
16	4	64	84	128	32
1024	10	10 240	11 274	20 480	5 120

Figure 5.4 Example to illustrate why $n + n \log n + \log n$ is $O(n \log n)$. Logarithms are to base 2 in all cases.

If we look at the numbers in Figure 5.4, we can see that for small values of n, the full cost of our search algorithm $(n + n \log n + \log n)$ is proportionally much greater than $n \log n$ – 2.5 times greater when n is 2. However, as n becomes large, the difference becomes proportionally much less important, so that when n is 1024, the full number of operations is only 10 per cent greater. Big O notation summarizes this situation by saying that if an operation is $O(n \log n)$, the number of operations it requires is always less than $n \log n$ times a constant factor, for all values of n above a certain value. Column 5 of Figure 5.4 shows the value of $2(n \log n)$. It can be seen that for all values of n above 2, this is greater than the full number of operations contained in the fourth column. In other words, in assessing the number of operations for a search, the function is dominated by the $n \log n$ term, except for small values of n. This means that in assesing complexity, factors which are either constant, or small relative to the main factor, are left out. This explains why in Section 5.1, the base of the logarithm was unimportant in reporting the complexity of a binary search. The reason is that a logarithm to one base can always be converted to a logarithm in another base by multiplying it by a constant factor (the logarithm of the first base to the second base in fact). For example, if m is $\log_2(n)$, then $\log_{10}(n) = \log_{10}(2) \cdot m$ where $\log_{10}(2)$ is simply 0.3010.

Figure 5.5 shows graphically how $n \log n$, $2(n \log n)$ and $n + n \log n + \log n$ vary as n increases. This makes it clear that $2n \log n$ is always greater than the total number of operations once n gets above about 3.

Figure 5.5 also makes it clear that big-O notation is concerned with the greatest amount of time an algorithm might take. It is also of interest to know whether there are situations in which the algorithm might need far fewer than $n \log n$ operations. In the case of the search operation, this might occur if the list of items was found to be sorted. However, checking for this

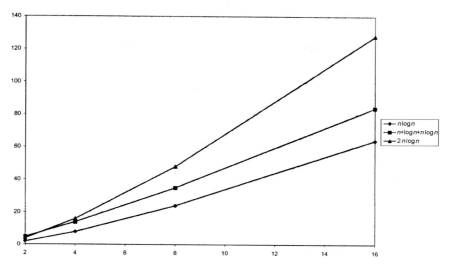

Figure 5.5 Graph of three functions of *n* from top to bottom: $2n \log n$, $n + n \log n + \log n$ and $n \log n$.

takes *n* operations, and doing the search takes $\log n$ so the efficiency in this case would still be $\Omega(n \log n)$. The Greek symbol Ω is used to indicate that this is a lower bound for this algorithm – in other words, it can never take less than c *n* log *n* operations, where c is a constant value. What about the search with unsorted data – can this ever do better than $O(n \log n)$? The answer is no, because as it stands it uses a sorting algorithm which will always take $n \log n$ operations. This is shown in Figure 5.4, which shows that if we multiply $n \log n$ by a different constant value (0.5 in this case), this will always be less than the number of full operations needed for the algorithm. This means that $n \log n$ describes both the upper and lower bounds for the algorithm, and this is expressed as $\Theta(n)$ using the Greek letter Θ.

Since this book is intended simply as an introduction to the ideas of algorithm design, we will not concern ourselves with the distinction between big-O, Ω and Θ notation, and will simply use big-O to describe the efficiency of different algorithms. This is quite common in the GIS literature, where the concern is normally with how well an algorithm will perform in the worst case, or in the average case, and so the lower complexity bounds are of less interest than the upper bounds.

FURTHER READING

Worboys (1995) contains a brief explanation of the measurement of algorithm complexity, and an introduction to big-O notation. Any introductory textbook on algorithms, such as Cormen *et al.* (1990), will contain a more extended discussion on the different forms of notation used for analysing efficiency. The NIST Dictionary of Algorithms and Data Structures (http://hissa.ncsl.nist.gov/~black/DADS/terms.html.) has a good description of big-O notation, with some examples, and links to the other related measures of complexity. Sorting is a classic computer science problem, and was the subject of one of the key publications in the Computer Science literature – Sorting and Searching by Donald Knuth. First published in the 1970s and recently revised (Knuth 1998) this is volume three in The Art of Computer Programming, a series which has been voted one of the twelve best scientific monographs of the century by Scientific American (ranking alongside the work of Einstein, non Neumann, Feynmann and Mandelbrot). Knuth is also the person who standardized the use of big-O notation. The NIST Dictionary page has lots of links to web pages that contain interactive demonstrations of the many sorting algorithms which have been developed.

6   Raster data structures

In Chapter 1, we saw that using a raster GIS we could store a set of spatial data in the form of a grid of pixels. Each pixel will hold a value which relates to some feature of interest at that point in space. These values are normally one of three possible types.

1   *Binary* – a value which indicates the presence or absence of a feature of interest. For example, in a layer representing roads, we might use 1 for pixels which contained part of a road, and 0 for pixels which did not.
2   *Enumeration* – a value from some classification. For example, a layer representing soils might contain codes representing the different soil types – 1 for Podsols, 2 for Brown Earths etc. Since the values are not directly related to the soil type, there would have to be a key of some sort indicating the meaning of each value.
3   *Numerical* – an integer or floating point number recording the value of a geographical phenomenon. In the soil example, we might have measurements of soil moisture content. A common example of this kind of raster layer is when the values represent the height of the land surface, in which case the layer is often referred to as a Digital Elevation Model (DEM) – these will be described in more detail in Chapter 9.

The raster data model has the great virtue of simplicity but it can produce very large files. The precision with which a raster layer can represent spatial data is related to the size of the pixel – you cannot represent anything which is smaller than a pixel. This means that pixel sizes need to be small, but the result is very large raster grids. For example, a single DEM tile from the British Ordnance Survey's PANORAMA dataset represents an area of 20 km by 20 km with pixels of 50 m – this is 160400 pixels (401 × 401) and only represents a small percentage of the land surface of Great Britain. What is more, for many applications, even smaller pixel sizes are desirable but halving the pixel size would increase the number of pixels by a factor of 4.

6.1 RASTER DATA STRUCTURES: THE ARRAY

The simplest method of storing a raster layer in the memory of the computer is using a data structure called an array. All programming languages have a structure called an array, which can be used to store and access lists of items. In Chapter 5, we considered alternative methods of searching through a list of entries in a telephone book to find one that matched with a particular name. The full list of names to be searched could be stored and accessed using an array as follows:

```
1   Array NAMES[1..64]
2   Read names from file into names array
3   i=1
4   FOUND=false
5   repeat until FOUND==true or i>64
6     if NAMES[I]=='Wise' then FOUND=true
7     i=i+1
```

This is the brute force algorithm for searching the list. The first line sets up an array with space for 64 names and the actual names are read from a file into this array. At this point, the first few elements of the array might contain the following:

NAMES [1]	Smith
NAMES [2]	Jones
NAMES [3]	Wise

Lines 5 to 8 go through this array, one item at a time, comparing the value with the name we are looking for – Wise – until either this is found, or the entire list has been searched. Each element in the array is identified by a number and this number is used to retrieve the correct element from memory. The array is available in programming languages because there are so many cases where it is necessary to deal with collections of related pieces of information. It is also possible to have arrays which have both rows and columns, and these are what can be used to store raster data.

The array is also an extremely efficient data storage mechanism but to understand why, it is necessary to understand something of the way a computer operates. Everyone is familiar with the idea of the storage of data on secondary media, such as floppy disks, zip disks and CD-ROMs. However, in order to be used by the computer, the data must first be transferred from the disk into the computer's memory. The only part of the computer that can actually do anything with the data is the Central Processing Unit or CPU. CPUs vary greatly in the way they are made and what they can do, but almost all have one thing in common – they can only deal with a few pieces of information at a time. This may seem surprising. How can a computer perform complex tasks, if it can only deal with a few things at a time? To see how it is able to do this, imagine being asked to work out the following sum on a calculator – 300 + ((25 × 320)/100). The steps involved would probably be:

1   Multiply 25 by 320 – this gives 8000.
2   Divide 8000 by 100 – this gives 80.
3   Add 80 and 300 which gives 380 – the answer.

Notice that in each step just three numbers are involved – two numbers which are input to the calculation, and one answer. It is always possible to break down problems in this way, into a series of steps which involve very few pieces of information, and this is exactly how computers are programmed to solve problems. The input data values are stored in the computer memory. From here the first two are passed to the CPU and operated on – 25 and 320 are multiplied to give 8000. This answer can be held in the CPU while the next input is fetched from memory. The 8000 is then divided by 100 to give 80. After the final step, the answer to the problem is passed back to the memory to be stored and allow the CPU to pass on to the next problem.

So why do we need memory? Why not simply pass the information directly between the secondary storage and the CPU? The answer is that this would be very slow, for three reasons. First, the transfer of information to and from disks along cables is inherently slower than transfer to and from memory which all takes place on printed circuit boards. Second, disks are mechanical devices which rotate, and this places an inherent limit on the speed with which information can be accessed. In contrast, memory works purely electronically. Third, finding an individual piece of information on a disk is relatively slow. Disks normally hold a large number of files, and once the correct file has been found, it is necessary to read through it to find the correct piece of data. In contrast, computer memory is designed to make it very easy to find individual data values.

Remember that each data value is held in one or more bytes of storage as we saw in Section 3.4. If we have a file containing a set of numbers, we can visualize these as being held in a series of boxes, one number per box. In a disk file the individual boxes are not normally distinguished – when the file is read, every box is retrieved starting with the first. In contrast, memory is organized so that every box has what is called an address, which is basically a number which uniquely identifies it. So our list of names might look like this in memory (Figure 6.1).

The circuitry in the computer is designed so that the information can be retrieved from any of the boxes equally quickly by passing the address to the

Address	12905	12906	12907	12908
Contents	Smith	Jones	Wise	

Figure 6.1 Storage of an array in computer memory.

CPU. It is rather as if the postal service worked by having a direct connection between every individual house and the post office.

What a computer program has to do therefore is work out the addresses of the boxes which contain the data which it needs. In the case of the array this is extremely simple. When an array is set up in a program, the program takes a note of the address of the first element – in this case the name 'Smith' which is stored in box 12905. The addresses of any of the other elements can then be worked out from the index number which is normally given in brackets after the name of the array. So when the program refers to NAMES[2] what the computer actually does is as follows:

1   The index value of this element is 2.
2   The first element in this array has an index value of 1.
3   This element is therefore $(2 - 1) = 1$ box on from the start.
4   The address of the first element is 12905.
5   The address of this element is therefore 12906.

This may seem long winded, but the computer only has to do two calculations – find how far along the array this element is (step 3) and use this to work out the actual address (step 5) – one subtraction and one addition. The calculation in step 3 produces what is sometimes called the offset – how far the element is from the start of the array. In many programming languages, the first array element is labelled 0 so that it is not necessary to perform the calculation in step 3 – the offset is simply the number of the element which can be added directly to the start address.

To see how this relates to GIS, let us consider the simple raster layer shown in Figure 6.2. Note that for clarity the pixel values are shown as letters which will help to distinguish them from the numerical memory

3	A	A	A	A
2	A	B	B	B
1	A	A	B	B
0	A	A	A	B
	0	1	2	3

Figure 6.2 Example of simple raster array.

addresses in the explanations which follow. In practice, most GIS systems only allow the storage of numerical data in pixels.

Instead of a list of names, we now have a set of rows and columns. When we want to identify a particular element in the array we will need to give both a row and column number – for instance IMAGE[3,3] to refer to the element in the top right hand corner. So does this mean we need a special form of memory which can handle 2D data and two sets of memory addresses – one for rows and one for columns? The answer is no in both cases – we still store our array in a sequence of memory locations in exactly the same way as for our list of names, as shown in Figure 6.3.

In order to do this we have to decide what sequence we will use to read the values from the rows and columns into memory. In this case, we have started in the bottom left hand corner, and proceeded from left to right along each row in turn until we reach the top. This will make the explanation of some of the other ways of storing raster data a little simpler, but in practice many GIS and Image Processing systems start at the top left and work their way down. There is no single agreed convention however and most GIS and Image Processing systems contain commands to 'flip' raster images which have been read in from systems which use a different convention.

So when a program refers to a particular pixel, such as IMAGE[2,3] how does the computer know which memory location to go to? The size of the array will have been stated at the start of the program. In our pseudo-code notation for example, a four by four array would be declared as follows:

```
Array IMAGE[0..3,0..3]
```

Notice that the rows and columns are both numbered from 0 to 3. It may seem more natural to number the rows and columns from 1, but in fact starting at zero makes some operations a little easier. As before, the program knows the address of the first item in the array – the pixel in the lower left hand corner. It also knows how many pixels are in each column. So if we count along two complete rows, and then three pixels along from the start of this row, we will be able to work out the address of the array element we want.

```
address=(nrow*rowsize)+ncolumn
address=(2*4)+3=11
```

Address	0	1	2	3	4	5	6	7	8	9	10	11	12	13	14	15
Value	A	A	A	B	A	A	B	B	A	B	B	B	A	A	A	A

Figure 6.3 Storage of array in computer memory.

You may like to check for yourself that IMAGE[2,3] is the 11th array element starting from the lower left hand corner.

Note that this calculation is not explicitly performed by the person writing a program in a language such as FORTRAN or C, who declares and uses arrays simply by putting the row and column positions in brackets after the name of the array. When the program is translated into an actual executable program by the compiler, one of the things which is done is to translate these statements into the sequence of operations which will calculate the address of the item and transfer it from memory to the CPU.

One important feature of this address calculation is that no matter how large the array held in memory, the retrieval of an item from it will take exactly the same amount of time. This is indicated by saying that the operation takes $O(1)$ time – the speed is the same, no matter how large the problem. Arrays can therefore be very efficient in terms of processing time. However, they are very inefficient in terms of storage, since every single pixel takes one element of storage. In order to assess the storage efficiency of various methods of handling raster arrays, it is easiest to think in terms of the number of rows or columns rather than the total number of pixels. For any given geographical area, this is determined by the resolution of the raster layer – halve the resolution and the number of rows and columns both double. However, the total number of pixels goes up by a factor of 4. This means that the array, which stores every pixel, has $O(n^2)$ storage efficiency, which is not very efficient at all. To solve this problem, various strategies have been adopted as we shall see in Section 6.2.

6.2   SAVING SPACE: RUN LENGTH ENCODING AND QUADTREES

The main disadvantage of the array is the size of the files when data is stored in this way. In the early days of GIS development, this was a serious problem. Even with modern computers with enormous amounts of disk space and memory, it still makes sense to reduce data sizes for a number of reasons. First, the transfer of data from disk to memory is considerably slower than the speed with which the same information can be processed once it is held in memory – therefore smaller files means quicker execution times. Second, the smaller the file size, the more images can be held in memory at one time. GIS analysis often involves viewing or using several layers – it is much slower if every time a new one is selected a file has to be moved out of memory to make way for it.

The simplest strategy for reducing file sizes is to use the smallest possible amount of storage for each pixel. We saw in Section 3.4 that for the storage of floating point numbers there is a need to store both a mantissa and an exponent, and as many digits as possible. For this reason floating point numbers are held in memory using at least 32 bits and often more. However, the same is not true of integers. An integer is held by converting the number from base 10 to base 2 and then storing the base 2 number. A single byte, with 8 bits, allows for a maximum integer of 255 as shown in Figure 6.4. If one of the bits is used to indicate the difference between positive and negative numbers then the range is from -128 to 127. Either of these is sufficient to hold the data in many raster layers, which often use small integer numbers – for instance Boolean layers only use the values 0 and 1 to indicate false and true respectively. Indeed these could be held using a single bit, but this is not normally an option which is available. However, the use of single byte integers is commonly available, and where appropriate will reduce the file size, and hence memory usage by a factor of 4 compared with using 32 bit words.

A second strategy for dealing with large files, is to hold only part of the layer in memory at any one time. In order to assess the efficiency of this approach, we have to consider two issues – how much memory will be needed, and how many times will we have to transfer data between memory and disk storage. Assume we have a layer of size n (i.e. with n^2 pixels in total). To process the whole array, we will have to transfer all n^2 pixels between the disk and memory, whether we copy them one at a time, or all at once. However, there is an extra overhead of time every time we ask for a transfer, because the system first has to find the location of the file on the disk, then find the particular part of the file we are requesting. Therefore, we need to try and minimize the number of times we go back and get extra data from the disk.

If we hold the whole array in memory then this uses $O(n^2)$ storage, but only requires 1 read and write operation between the disk and the memory. At the other extreme, we could read each pixel as we need it and write it back to disk afterwards. This now uses 1 unit of storage but $O(n^2)$ read/write operations. The first option is very quick, but uses a lot of memory – the second uses almost no memory but would be painfully slow. A compromise is to read one row at a time into memory, process it and write it out to disk – this uses $O(n)$ storage, and also $O(n)$ read/write operations. The difference between these approaches can be quite marked. Wise (1995) describes an example where this was a real issue. The work was on the

Binary	Decimal
00000000	0
00000001	1
11111111	255

Figure 6.4 Examples of storage of integers in bytes.

problem of capturing raster data from scanned thematic maps, such as soil or geology maps. As part of this a program was written which processed a scanned image, replacing pixels which represented things like text labels, lines etc. with the value for the likely soil or geology category at that point. The program was written for what was then the latest version of the IDRISI GIS, which worked under MS-DOS, and could therefore only access 640 Kb of memory. Even with nothing else stored in the memory, the largest image size which could be held in memory would only have been just over 800 columns by 800 rows – in contrast, by processing a row at a time images of up to 640000 columns could be handled, with no limit on the number of rows.

These strategies may help, but there are also other things we can do in order to reduce the size of the image which needs to be stored on disk or held in memory. Section 6.1 described each of the three main types of values stored in raster GIS layers – binary, enumerated and numerical. In the case of the first two, because the features we are representing occupy regions of the map, the raster layers contain large number of pixels with the same value next to one another. We can exploit this characteristic to save storage space and the simplest way to do this is to use what is called run length encoding. Consider the simple raster layer we used in Section 6.1 which is repeated as shown in Figure 6.5.

When we stored this as a full array, the first 3 pixels all contained the same value – A. What we have is a sequence or run of pixels, and instead of storing each one we can store the information about the run – how long it is and what value the pixels have. Applying this to the whole layer produces the result shown in Figure 6.6. Even with this small example we have reduced the number of bytes of storage used for the layer. But will we

3	A	A	A	A
2	A	B	B	B
1	A	A	B	B
0	A	A	A	B
	0	1	2	3

Figure 6.5 Example of simple raster array.

Address	0	1	2	3	4	5	6	7	8	9	10	11	12	13	14	15
Value	3	A	1	B	2	A	2	B	1	A	3	B	4	A		

Figure 6.6 Storage of a run length encoded layer in computer memory.

always save space in this way? The answer unfortunately is no. Imagine a layer in which every pixel was different from its neighbours, such as a DEM. Every pixel would take 2 bytes of storage instead of $1 - 1$ to record a run length of 1, and one for the value itself – so the file size would double.

The final raster data structure we will consider is called the quadtree, and it extends the idea of run length encoding to 2D. If we look at Figure 6.5, we can see that there is a block of 4 pixels in the lower left hand corner which all have the value A. Instead of storing four small pixels, it would be far more efficient to store 1 pixel, which was four times the size of the 'normal' pixel. This is the basis of the quadtree method in which the pixel size is allowed to vary across the image, so that uniform areas are stored using a few large pixels, but small pixels are used in areas of variation. To illustrate how this works, let us apply it to the layer shown in Figure 6.5. At the first stage, the layer is divided into four quadrants, as shown in Figure 6.7. Each quadrant is numbered from 0 to 3 in the sequence shown, which is known as Morton order after its inventor. The reason for this particular sequence will become clear later. If we examine each quadrant, we can see that quadrant 0 does not need to be subdivided any further – the values in all the pixels are the same, and so our new pixel 0 can be used to store this data. The three other quadrants are not uniform and so must be subdivided again. Notice that each of the new pixels is labelled by adding a second digit, also in Morton order, to the digit from the first level of subdivision.

Because this is only a 4 by 4 image, the process stops at this point. But how do we store this information in memory, especially now that the pixels are no longer the same size? One method is shown in Figure 6.8.

The first four memory locations (with addresses 0 to 3) are used to store the results of the quadrants from the first subdivision of the image. The first quadrant, labelled 0, was uniform, and so we can store the pixel value – A. The second quadrant, labelled 1, was not uniform, and so we are going to need four bytes to store whatever we found when we subdivided this quadrant. The next available location is at address 4, so we store this address in location 1. Since this address is pointing to the location of

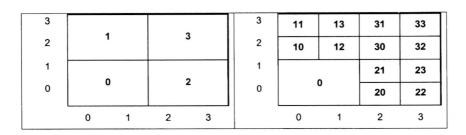

Figure 6.7 Quadtree subdivision of layer shown in Figure 6.5.

Address	0	1	2	3	4	5	6	7	8	9	10	11	12	13	14	15
Quadtree address	0	1	2	3	10	11	12	13	20	21	22	23	30	31	32	33
Value	A	4	8	12	A	A	B	A	A	B	B	B	B	A	B	A

Figure 6.8 Storage of quadtree in memory.

another piece of information, it is known as a pointer. We have to do the same thing for quadrants 2 and 3, storing pointers to addresses 8 and 12 respectively. The four address locations starting at 4 are used to store the results of subdividing quadrant 1 to produce 10, 11, 12 and 13 – since these were all uniform, we can simply store the pixel values, and in fact this is true for all the remaining pixels.

In this case, we have not saved any space at all compared with the original array method, because there are not enough large uniform areas on the layer. In fact, as with run length encoding, we could end up storing more information. If quadrant 0 had not been uniform, we would have needed an extra four bytes of storage to store the individual pixel values making up this quarter of the image. However, in real world examples the savings in space can be considerable.

In this example, the image was conveniently the right size to make it possible to divide it into four equal quadrants. When this is not the case (i.e. most of the time) the image is expanded until its sides are both powers of 2, filling the extra pixels with a special value indicating 'no data'. This increases the amount of storage of course, but since the extra pixels are all the same, they can generally be represented using fairly large pixels, and the additional data are more than offset by the savings due to the quadtree itself.

So why is it called a quadtree? The quad part is obvious, from the subdivision into quadrants. The tree comes from a common way of representing such data structures in Computer Science as shown in Figure 6.9.

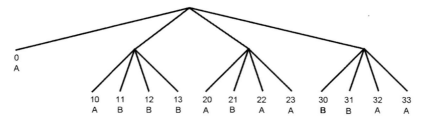

Figure 6.9 Graphical representation of quadtree.

The first level of subdivision is represented as four branches from the original image. Where quadrants are subdivided, then four further branches are drawn, giving a tree-like structure. The ends of the lines are all called nodes – those where the process ends (as in the case of 0 and 10, 11, 12 and 13 etc. in the diagram) are called leaf nodes, while those which represent points of further subdivision are called branch nodes. The first node of the tree is called the root node, even though trees are usually drawn with their origin at the top, as in Figure 6.9.

In Figure 6.8, the Morton addresses of the pixels have been shown, for clarity. In fact, this information is not stored in reality because there is a simple method for calculating it from the row and column number of a pixel as we will see in Chapter 7, when we look at how we can use these different data structures to answer a range of GIS queries.

FURTHER READING

Brian Berry's two books (Berry 1993; 1995), based on his articles in GIS World, provide a gentle introduction to raster data structures and analysis. The NCGIA Core Curriculum Units 35 and 36 (http://www.geog.ubc.ca/courses/klink/gis.notes/ncgia/toc.html) cover raster data structures for raster data. The classic texts on quadtrees are the two books by Samet (1990a,b). These describe a range of different quadtree data structures and algorithms for performing standard GIS operations such as overlay using quadtrees. There is now also a web site (http://www.cs.umd.edu/~brabec/quadtree/) by Brabec and Samet which provides applets demonstrating many quadtree algorithms. Try one of the point or rectangle quadtree demos – add some points/rectangles, and then move one to see how the quadtree decomposition of space changes to store the information as efficiently as possible. Rosenfeld (1980) and Gahegan (1989) are both shorter summaries which provide a good overview, while Waugh (1986) presents some cogent arguments as to why quadtrees may not be a good thing after all!

7 Raster algorithms

The previous chapter described some of the commonest data structures for the storage of raster data and emphasized that at least one of the reasons that they were developed was to produce smaller files for raster GIS layers. Smaller files mean that less disk space is needed to store them and that it will be quicker to transfer them between the disk and the memory of the computer. A small file size should also mean that operations on the layers, such as queries and overlays, should run more quickly, because there is less data to process. However, this will only be true if the query can use the raster layer in its compressed form, whether this be run length encoded or in a quadtree or in any other format. If this is not possible then the layer would have to be converted back to its original array format (which would take time) and then the query run on this expanded form of the data. So the data structures which are used for raster data not only have to produce savings in file size, they must be capable of being used for a range of GIS queries. To illustrate this, we will consider some simple raster GIS algorithms, and show how these can be implemented using the three raster data structures described in Chapter 6.

7.1 RASTER ALGORITHMS: ATTRIBUTE QUERY FOR RUN LENGTH ENCODED DATA

The first is the most basic query of all – reporting the value in a single pixel. To illustrate this we will use the simple layer in Figure 7.1, and imagine that we wish to find out the attribute value stored in the pixel in row 2 column 3. Remember that rows and columns are numbered from zero.

In a real GIS of course, this query is more likely to take the form of an on-screen query, in which the user clicks a graphics cursor over the required pixel on a map produced from the data. However, it is a simple matter for the software to translate the position of the cursor to a pixel position on the layer itself.

We have already seen how this query would be answered using the array data structure in Section 6.1. Assuming the layer is stored in an array called

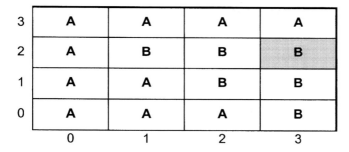

Figure 7.1 Example of simple raster array – pixel in row 2 column 3 is highlighted.

IMAGE, the value we want will be held in the array element IMAGE[2,3]. To retrieve this value from the array as it is stored in memory (Figure 7.2) we calculate the address from the row and column number by first calculating the offset – how many elements along the array we need to go:

```
offset=(nrow*rowsize)+ncolumn
offset=(2*4)+3=11
```

To find the actual address in memory, we add the offset to the address of the first element, which is 200

```
address=origin+offset=200+11=211
```

If we look at Figure 7.2, we can see that the value in address 211 is B, which is correct.

Notice that the addresses are different from those given in Chapter 6. This is deliberate because the actual addresses will normally be different every time the software is run. When a program, such as a GIS, is loaded into memory ready to be run, where exactly in memory it goes will depend on what is already running on the computer. If this is the first package to be run, then it will be loaded near the start of the memory (and the addresses will be small numbers). If there are already other programs running, such as a word processor or spreadsheet package, then the GIS will go higher in memory and the addresses will be larger numbers. One of the things which

Address	200	201	202	203	204	205	206	207	208	209	210	211	212	213	214	215
Value	A	A	A	B	A	A	B	B	A	B	B	B	A	A	A	A

Figure 7.2 Storage of array in computer memory.

happens when a program is loaded into memory, is that the origins of all the arrays are discovered and stored, so that all the array referencing in the program will work properly.

If we look at the same raster layer stored in run length encoded form (Figure 7.3), we can see that this same approach is not going to work because there is no longer a simple relationship between the addresses and the original row and column numbers. For instance, the value stored at offset 4 will always refer to the second run found in the original layer but which row and column this relates to depends entirely on the length of the first run. What we have to do is work out the offset ourselves, and then count along the runs until we find the one that contains the pixel we want.

The program we use to do this will be as follows:

```
1   array RLE[0..15]
2   offset=(nrow*rowsize)+ncolumn
3   pos=0
4   n=0
5   repeat until pos>=offset
6     pos=pos +RLE[n]
7     n=n+2
8   value=RLE[n−1]
```

The Run length Encoded storage of the image is declared as a one-dimensional array on line 1. Since we cannot take advantage of the automatic facilities provided by the 2D row and column notation, we may as well treat the image in the way that it is actually stored in memory – as a single row of values. Line 2 calculates the offset, and is the same as the calculation that is done automatically when the image is stored using a 2D array. The key to this algorithm is line 6. The first time through this program, both pos and n will have the value 0 (set on lines 3 and 4). RLE[0] contains the length of the first run (3) so pos will take the value 3. If this is greater than or equal to the offset we are looking for (line 8), then this first run would contain the pixel we are looking for. However it does not, so we go back to the start of the repeat loop on line 5. When line 6 is executed this time, n has been set to a value of 2 (on line 7) so RLE[2] will give the length of the second run – 1. This is added to pos, making 4, which is still less than our offset, so we go round the repeat loop again. Eventually, on the sixth time around, the value of pos will reach 12 – this means we have found the run we are

Address	200	201	202	203	204	205	206	207	208	209	210	211	212	213	214	215
Value	3	A	1	B	2	A	2	B	1	A	3	B	4	A		

Figure 7.3 Storage of image in run length encoded form in computer memory.

looking for and the repeat loop ends. By this stage, n has a value of 4, since it is pointing to the length of the next run, so we look at the value in the previous array element to get the answer – B. It is immediately clear that this algorithm is less efficient than the one for the array. In fact, every time we make a reference to an element in RLE in this example, the computer will have to do the same amount of work as it did to retrieve the correct answer when we stored the data in an array. Answering this query using an array had order $O(1)$ time efficiency – it works equally quickly, no matter how large the raster layer. In the case of the run length encoded data structure, the time taken is related to the number of runs which are stored. Other things being equal, we might expect this to be related to the size of the layer. The layer in 7.1.1 is 4×4 pixels. If we doubled the size to 8×8, the number of pixels would go up by a factor of 4. However, the number of runs would not go up quite this much, because some runs would simply become longer. The exact relationship would depend on how homogeneous the pattern of values was. A reasonable assumption is that the runs would roughly double in size if n doubled – hence our algorithm would have $O(n)$ processing efficiency.

7.2 RASTER ALGORITHMS: ATTRIBUTE QUERY FOR QUADTREES

We seem to face the same problem in carrying out our attribute query using a quadtree as we did with the run length encoded version of the layer – there appears to be no relationship between the position of the pixels in the original GIS layer, and their position in the data structure (Figure 7.5). However, oddly enough, this is not the case, because when we created the quadtree, we saw that we were able to give each 'pixel' in the tree a Morton address (shown on Figure 7.4), and there is a relationship between these Morton addresses and the row and column numbers.

Let us take the same query as before – find the value of the pixel in column 3, row 2. To discover the Morton address we first have to convert

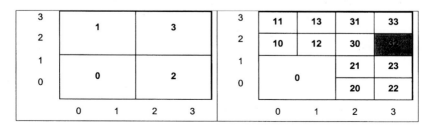

Figure 7.4 Quadtree subdivision of layer shown in Figure 6.5, with the pixel in column 3, row 2 highlighted.

the column and row to binary. For those who are not familiar with changing between number bases, this is explained in detail in a separate box – those who are familiar will know that the result is as follows:

$$3_{10} = 11_2$$
$$2_{10} = 10_2$$

where the subscript denotes the base of the numbers. We now have two, 2-bit numbers – 11, and 10. We produce a single number from these by a process called interleaving. We take the first (leftmost) bit of the first number followed by the first bit of the second number, then the second bit of the first number and the second bit of the second number as shown below:

First bit of 11	First bit of 10	Second bit of 11	Second bit of 10
1	1	1	0

This produces a new number 1110_2. Finally, we convert this from base 2 to base 4 which gives 32_4. Again this is explained separately in the box for those who are unfamiliar with conversion between different number bases. If you look at Figure 7.4, you can see that this is indeed the Morton address for the pixel in column 3, row 2. So why does this apparently odd process produce the right answer? The answer lies in the way that the Morton addresses were derived in the first place.

One of the problems with spatial data is just that – that they are spatial. One of the most important things to be able to do with any kind of data is to put it into a logical sequence, because it makes it easier to find particular items. We have already seen how having a list of names in alphabetical order makes it possible to find a particular name using things like a binary search. This algorithm is only possible because the names are in a sequence. But how can we do the same with spatial data, which is 2D? If we take pixels and order them by their row number, then pixels which are actually very close in space (such as row 0, column 0 and row 1, column 0) will not be close in our ordered list. If we use the column number instead we will have the same difficulty. This was the problem that faced the developers of the Canadian GIS, and the solution they came up with was to combine the row and column numbers into a single value in such a way that pixels that were close in space, were also close when ordered by this value. The Morton code, named after its inventor, was the result. Therefore, the reason that quadtree pixels are labelled in this way is because there is then a simple means of deriving their quadtree address from the row and column number. The Morton code is also known as a Peano key, and is just one of a number of ways in which pixels can be ordered. The question of how spatial data can be organized to make it easy to search for particular items is a very important one, and will be considered further in Chapter 8.

Offset	0	1	2	3	4	5	6	7	8	9	10	11	12	13	14	15
Quadtree address	0	1	2	3	10	11	12	13	20	21	22	23	30	31	32	33
Value	A	4	8	12	A	A	B	A	A	B	B	B	B	A	B	A

Figure 7.5 Storage of quadtree in memory.

Let us return to our original query. The next stage is to use our Morton address of 31 to retrieve the correct value from the data structure in Figure 7.5. Each digit in the Morton address actually refers to the quadrant the pixel was in at each level of subdivision of the original layer, and this is the key to using the address in conjunction with the data structure. In this case, the first digit is 3, which means the pixel was in quadrant 3 when the layer was first subdivided (this can be checked by looking at Figure 7.4). When we stored the quadtree in our data structure, the results for the first subdivision were stored in the first four locations, starting at offset 0. (Note that to simplify this explanation, the storage elements in Figure 7.5 are shown in terms of an offset, rather than a full address in memory as was done for the example of run length encoding). The value for pixel 3 will therefore be stored in offset 3 – the first digit of the Morton address points directly to the location we need. The value stored in offset 3 is number 12. Since this is a number, rather than a letter, it means that the results of the second subdivision are stored in the four locations starting at offset 12. The second digit of our Morton address is a 2, so the value we are looking for will be stored in the location which is offset from 12 by 2 – a new offset value of 14. Looking at Figure 7.5, we can see that the value stored at offset 14 is indeed B, the answer to our query.

What would have happened if we had requested the value for the pixel on row 0, column 1? The Morton address in this case would be 0001_2, or 01_4. Note that we do not leave off the leading zeroes from these addresses. Applying the same algorithm as before, we take the first digit, 0 and look in our data structure. This time we find a letter, A, rather than a number. This means that we have found a leaf, rather than a branch, and in fact this is the answer to our query. There is no need to use the other digits in the Morton address, since all the pixels in quadrant 0 at the first subdivision had the same value.

The pseudo-code to find a value from the data structure given the Morton code would look something like this:

```
1  Array QUADTREE[0..15]
2  found=false
```

```
3   repeat until found==true
4     Take next digit from address
5     offset=offset+digit
6     if QUADTREE[offset] is a leaf then
7       value=QUADTREE [offset]
8       found=true
9     else
10       offset=offset+QUADTREE[offset]
```

As with the run length encoded version, we treat the quadtree as a single array, and find our way to the correct element of the array using an algorithm rather than the in-built system which we had with the 2D array. The efficiency of this algorithm will be related to the number of times we have to go round the repeat loop in the program above. Each time we go round this loop we are going down one further level of the tree, or one further level of subdivision of the original image. In our simple example, we had a 4×4 image, which was divided into quadrants two times. If we doubled the size of the image to 8×8, this would only add one to the number of subdivisions, which would become 3. You will probably have noticed a pattern here, since

$$4 = 2^2$$
$$8 = 2^3$$

In other words, the maximum number of levels in a quadtree is the number of times we can divide the size of the quadtree side, n, by 2 – in other words $\log_2(n)$. This means our algorithm for finding any given pixel has processing efficiency which is $O(\log n)$.

Before we leave this particular algorithm, it is worth taking time to explore the fourth line in the code sequence above since it affords a useful insight into the way that computers actually perform calculations. Given the Morton address 31, the first time through the repeat loop we need to select digit 3, and then digit 1 the next time around. But how is this done? Imagine for a moment that we needed to do the same thing in the familiar word of base 10 arithmetic. To find out what the first digit of 31 was we would divide the number by 10, throwing away the remainder: $31/10 = 3.1 = 3$ with remainder 0.1. When we divide by 10 all we do is move the digits along to the right one place – if we multiply by 10, we move them to the left and add a zero at the right. Exactly, the same principle applies with binary numbers.

In the memory of the computer the address 31_4 would be stored as a 32 bit binary number:

00000000000000000000000000001101   [1]

Notice that 2 binary digits (bits) are needed for each of our base 4 digits – 3 is 11, 1 is 01. If we move all the bits two positions to the right, we will get the following:

00000000000000000000000000000011   [2]

This is our first digit 3, which is what we need. We have divided by 100_2 but we have done it by simply shifting the bits in our word, an operation which is extremely simple and fast on a computer. In fact on most computers this sort of operation is the fastest thing which can be done. The second digit we need is 1, which looks like this:

00000000000000000000000000000001   [3]

The bits we need are already in the correct place in the original Morton address [1] – all we need to do is get rid of the 11 in front of them. The first step is to take the word containing the value 3 [2], and move the bits back two positions to the left giving us the following:

00000000000000000000000000001100   [4]

This moves the 11 back to where it started, but leaves zeroes to the right of it. If we now subtract this from the original Morton address [1], we will be left with 01.

This approach will work no matter how large the Morton address. Assume we have an address which is 1032_4. For simplicity, we will assume this is stored in an 8 bit word, rather than a full 32 bit word. The algorithm for extracting the leftmost digit is as follows:

Step	Result
Store Morton address in R1 and make an exact copy in R2	R1: 01001110 R2: 01001110
Shift R1 right by 6 bits. R1 now contains first digit	R1: 00000001 R2: 01001110
Shift R1 left by 6 bits	R1: 01000000 R2: 01001110
Subtract R1 from R2	R1: 01000000 R2: 00001110
Copy R2 to R1. Now ready for extraction of second digit	R1: 00001110 R2: 00001110

At the end of this, R1 contains the original Morton address, but with the first digit (the first two bits) set to zero. To obtain the second digit, we can

Conversion between base 10 and base 2

If we have a number in our usual base 10 numbering system, how do we convert this to a different system, such as binary? First, it is important to understand that any number, in any number base, is effectively a sum. Imagine trying to count a flock of 27 sheeps on your hands. When you get to 10 you run out of fingers, so what do you do? You might use your toes instead but this will only allow you to get to 20 before you run out of things to count with. Alternatively you might make a mental note that you have counted 1 lot of 10 and start again. The second time through the fingers you get to 20, make a mental note that this is the second lot of 10 and start again until you get to the last sheep. Your count is therefore 'Two lots of ten plus seven more' which is 27 in total. In other words, the number 27 can be thought of as:

$$27 = 2 \cdot 10 + 7$$

Instead of fingers of course we use symbols to represent numbers – the familiar digits from 0 to 9. The same logic can be applied to counting our flock of sheep. We count up to 9. At this point, we have run out of digits. We could keep inventing new symbols for larger numbers but this would be very cumbersome. Instead we record the 10th sheep using the digit 1 – 1 lot of 10, and start counting from 1 again. If the number of sheep were 271, this method runs out of digits again at 99 – we don't have another digit to record that we have reached the tenth lot of tens. So we use 1 to record that we have reached 10 lots of 10 for this first time. Therefore, the number 271 can be represented as:

$$271 = 2 \cdot 10 \cdot 10 + 7 \cdot 10 + 1$$

Ten times ten can also be represented as 10^2 (ten squared). Ten itself is 10^1 (ten to the power 1) since any number to the power 1 is simply itself. One is 10^0, since any number to the power nought is 1. So our number becomes a sum of successive powers of 10:

$$271 = (2 \cdot 10^2) + (7 \cdot 10^1) + (1 \cdot 10^0)$$

In our decimal numbering system, powers of 10 are used because there are 10 digits (0–9) and because the system originated from counting on our 10 fingers (which are also called digits of course – hence the name for the mathematical digits). However, we do not have to use 10 as our number base. When designing computers, it was found that it was far easier to design machines which could use just two digits – 0 and 1 – for storing and processing information. For instance, simple switches could be used, since they have just two states – on and off. Storing numbers in base 2 is exactly the same as in base 10 – for instance the number 11011 is 27 in binary:

$$11011 = (1 \cdot 2^4) + (1 \cdot 2^3) + (0 \cdot 2^2) + (1 \cdot 2^1) + (1 \cdot 2^0)$$
$$= 16 + 8 + 0 + 2 + 1$$
$$= 27$$

This provides us with a method of converting between numbers in different bases. To convert from binary to decimal, we make a table of what the successive powers of 2 are in the decimal system. So $2^0 = 1$, $2^2 = 4$ and so on. In theory, we could use the same approach to convert from decimal to binary, by making a table of what the various powers of 10 are in binary, multiplying them by the value of the relevant digit (also converted to binary) and summing the result (also in binary). An alternative method which allows us to work in our familiar base 10 arithmetic, is to take the base 10 number and successively divide it by the new base of 2. At each stage, we note whether the remainder is 0 or 1 and this will be one of the digits in our binary number. We continue this process until we can no longer divide by 2. With 27 the process looks like this:

$$\frac{27}{2} = 13 \text{ remainder } 1$$
$$\frac{13}{2} = 6 \text{ remainder } 1$$
$$\frac{6}{2} = 3 \text{ remainder } 0$$
$$\frac{3}{2} = 1 \text{ remainder } 1$$
$$\frac{1}{2} = 0 \text{ remainder } 1$$

Notice that the remainders, read from top to bottom, give us our number in base 2.

Conversion between base 2 and base 4

Conversions between base 2 and base 4 are much simpler than the general case because of the fact that four is simply two squared. This means that every base 4 digit, is represented by exactly two base 2 digits as follows:

Base 4	Base 2
3	11
2	10
1	01
0	00

To convert a number from base 2 to base 4, all you have to do is group the base 2 digits in pairs, starting at the right, and replace each pair with the equivalent base 4 digit. So to convert 11011 to base 4:

$$11011 = 01\ 10\ 11 = 123_4$$

Note that the number has to be padded out to an even number of binary digits on the left, so the leftmost 1 becomes 01. Conversion from base 4 to base 2 is simply the reverse process – replace each base 4 digit with the appropriate pair of base 2 digits.

The same principle applies to conversion between any number bases which are powers of 2. One which is frequently used is the conversion of binary numbers to base 16, or hexadecimal. Since base 16 has more than 9 digits, letters A–F are used for the extra digits. So the conversion table between binary and hexadecimal is as follows:

Decimal	Binary	Hexadecimal	Decimal	Binary	Hexadecimal
0	0000	0	8	1000	8
1	0001	1	9	1001	9
2	0010	2	10	1010	A
3	0011	3	11	1011	B
4	0100	4	12	1100	C
5	0101	5	13	1101	D
6	0110	6	14	1110	E
7	0111	7	15	1111	F

The advantage of using hexadecimal is that it provides a convenient way of representing long binary numbers. For instance, a 32 bit word can be represented using 8 hexadecimal digits instead of 32 binary digits. Error messages which report memory addresses or values commonly use hexadecimal for this reason.

simply repeat this sequence, but shifting by 4 bits rather than 6. Then for the third digit, we shift by 2 digits, and the final digit by 0 digits (i.e. we do not need to do anything). The size of the original address was 4 base-4 digits. Given a Morton address of size n, this will be stored in $2n$ bits. To get the first Morton digit we shift by $2n - 2$, to get the second by $2n - 4$ and so on – so to get the mth digit we shift by $2n - 2m$ bits.

The reason that the variables above are called R1 and R2 is that in reality this sort of operation is done by storing the values in what are called registers. These are special storage units which are actually part of the CPU rather than part of the memory of the computer, and the CPU can perform basic operations, such as addition, subtraction and bit shifting on them extremely quickly.

We have assumed that we already have the Morton address we need, but of course we will need to work this out from the original row and column numbers. Since these will be stored as binary integers they will already be in base 2. In order to interleave the bits, we will also use a series of bit shifting operations to extract each bit from each number in turn, and add it in its place on the new Morton address. As with extracting the digits from the Morton address, it may seem very complicated, but all the operations will work extremely quickly on the computer.

7.3   RASTER ALGORITHMS: AREA CALCULATIONS

The second algorithm we will consider is one which we considered in the context of vector GIS in Chapter 4 – calculating the size of an area. In vector, the procedure is quite complicated, because areas are defined as complex geometrical shapes. In principle, the raster method is relatively straightforward. To find the area of the feature labelled A (shaded area in Figure 7.6) we simply count up the number of A pixels. As we shall see things are more complicated when we use a quadtree, but let us start with our simplest data structure – the array.

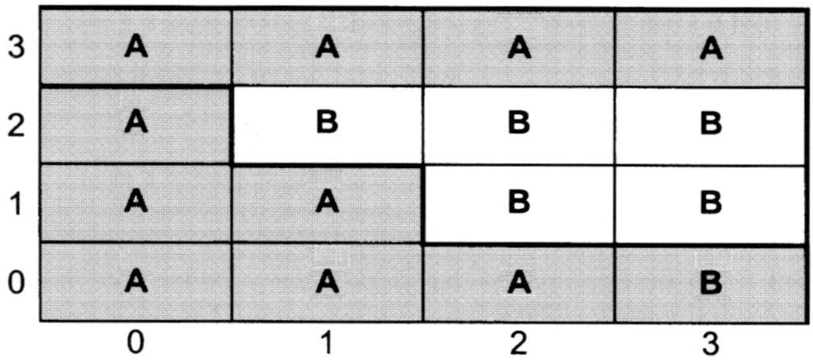

Figure 7.6 Example raster layer.

The algorithm to calculate the size of the shaded area A is pretty much the same as the description above:

```
1  Array IMAGE[0..maxcol,0..maxrow]
2  area=0
3  for i=0 to maxrow
4    for j=0 to maxcol
5      if image[i,j]='A' then area=area+pixelsize
```

The program loops through each column in each row, increasing the value of the variable `area` by the size of one pixel each time an A pixel is found.

The program for the run length encoded data structure (Figure 7.7) is actually not very different. Each run already contains a count of the number of pixels, so for each run of type A, we simply use this to update our area variable.

```
1  area = 0
2  for i=1 to nruns
3    if run.type=='A' then
4      area=area+pixelsize*run.length
```

It is easy to see that the area calculation algorithm has $O(n^2)$ complexity using a 2D array but $O(n)$ complexity with a run length encoded image (assuming that the number of runs is linearly related to the layer size). In this case, the run length encoded data structure can actually process this query more efficiently than the simple array, in contrast to the pixel value query.

Before moving on to look at the quadtree algorithm, it is worth looking at the algorithm for the run length encoded structure to illustrate a number of points about the way in which algorithms are actually implemented in practice. In order to implement this algorithm we would need some way of storing and accessing the run length encoded image. One method is to use a one-dimensional array, and step through this two elements at a time:

```
1  Array RLE[0..nruns*2]
2  for i=1 to nruns step 2
3    if RLE[i]=='A' then
4      area=area+pixelsize*RLE[i+1]
```

Address	0	1	2	3	4	5	6	7	8	9	10	11	12	13	14	15
Value	3	A	1	B	2	A	2	B	1	A	3	B	4	A		

Figure 7.7 Storage of layer in run length encoded form.

Some languages contain the facilities for more complex data structures than an array, and would allow us to directly declare a structure which contained two elements. For instance in C, the following could be used:

```
struct rle {
  integer:runlength;
  integer:runtype}
rle image(5);
```

This sets up an array called image where each element of the array is a pair of items – the run length and the pixel value associated with it. Using this structure, something like our original pseudo-code could be used in which we loop through each run, referring to both its length and type as required. The pseudo-code assumed that we knew how many runs there were in the image – but how would this information actually be stored? In practice, this could be handled in a number of ways. One would be to alter the data structure, to put a count of the number of runs at the start, as shown in the upper example of Figure 7.8. The program would then read this as the value of nruns. The second, which avoids having to count up the number of runs, is to put what is sometimes called a 'flag' at the end of the data – in the lower figure of Figure 7.8 this is a negative run length – which signals the end of the data. A final point is that this algorithm would be more likely to be programmed as shown below:

```
1  area=0
2  npixels=0
3  for I=1 to nruns
4    if run.type=='A' then
5       npixels=npixels+run.length
6  area=npixels*pixelsize
```

Address	200	201	202	203	204	205	206	207	208	209	210
Value	5	6	A	2	B	1	A	3	B	4	A

Address	200	201	202	203	204	205	206	207	208	209	210
Value	6	A	2	B	1	A	3	B	4	A	−1

Figure 7.8 Two techniques for keeping track of the number of runs. Upper: count at the start of the data. Lower: negative 'flag' value to signal the end of the data.

The difference is that rather than keeping a tally of the area, as we did for the full array, we keep a tally of the number of pixels (line 5) and then multiply this by the pixel size at the end (line 6). The reason is that addition is a much quicker operation on the computer than multiplication. In this second version of the program we perform as many additions as there are runs, but only one multiplication – the first version performed an addition and a multiplication for every run. In overall terms both algorithms have $O(n)$ complexity, where n is the size of the raster layer. However, we can also consider the processing efficiency of the details of how the algorithm is coded. Even if we assume that multiplication and addition run at the same speed, then the first algorithm will need $2n$ operations where n is the number of runs in the image, while the second will need $n + 1$. If we are dealing with an image with several hundred runs, then a change which at least halves the running time is worth having. This simple example illustrates that the analysis of efficiency can be applied to all levels of algorithm design, from the overall approach down to the slightest detail.

Let us now consider how we perform the area query using the quadtree data structure. Counting pixels will no longer work, since we have replaced the original pixels with new ones of different sizes. If we look at the data structure itself, we can see that it would be an easy matter to go through counting the cells labelled A, but there is nothing which explicitly tells us how large each A is. This information could be stored of course, but it would increase the size of the quadtree, since each group of four elements in the structure would have to have a code indicating how large any leaf nodes in that particular group of four were.

A solution which avoids having to store the information is to start at the top of the tree, and visit each leaf in turn, keeping track of what size it is. To explain the principle, Figure 7.10 illustrates the quadtree shown in Figure 7.9, but in diagrammatic tree format, and showing the values at the leaf nodes.

We start at the root node of the tree. In Figure 7.10, this is actually the node at the top of the diagram but since it represents the original image, which is then subdivided, it is known as the root of the quadtree. We know the area of the root node, since it is the area of the original layer – 16 (assuming for simplicity that each pixel has an area of 1). We visit the first node in the tree which is a leaf node with a value of A. We have come down one level so the node area is now $16/4 = 4$. We therefore add 4 to our tally

Offset	0	1	2	3	4	5	6	7	8	9	10	11	12	13	14	15
Value	A	4	8	12	A	B	B	B	A	B	A	A	B	B	A	A

Figure 7.9 Storage of layer in linear quadtree form.

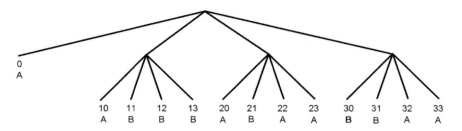

Figure 7.10 Quadtree representation raster image from Figure 7.6.

of area and continue. We continue by going back up the tree, and coming down to the next node. The area is again 16/4, but this time we have reached a branch. We therefore go down the first node from this branch, which will have an area of 4/4 = 1. This is a leaf, which has a value of A, so we add 1 to the area tally. We then go back up, and down the next branch – this is a leaf but has a value of B so we ignore it and carry on. We continue this process, until we have traversed every branch of the tree in sequence, by which time we will have our answer.

This is fine diagrammatically, but how would we program this algorithm in our computer? Let us start by considering the four cells labelled ABBB in Figure 7.10. To add up the area of A pixels for these four leaves we could have a program which looked as follows:

```
leafarea=1
for node=0 to 3
   if value=='A' area=area+leafarea
```

In other words – visit each leaf in turn and if its value is 'A' add the value of `leafarea` to the total. Now consider the four branches coming down from the root of the tree. These can be dealt with using the same approach, but with one difference – some of these nodes are not leaves but branches, so we will need to deal with them differently. Let us first modify the program to include the test for whether a node is a leaf or a branch:

```
1  leafarea=4
2  for node=0 to 3
3     if node is a leaf then
4        if value=='A' area=area+leafsize
5     else /* node is a branch
6        Count 'A' leaves under branch
```

If we find a node is actually a branch, then we will need to have a program which will visit the four nodes under this branch, checking whether they are leaves with value A and if so adding up their area. But this is exactly what the program we are currently writing does! Does this mean that we have to write a second program, exactly the same as this, which can be called for nodes at the second level of the tree? Fortunately the answer is no – all we do is call the program from within itself, as shown in the full algorithm below.

```
1   procedure area(leafsize)
2   local leafarea=leafsize/4
3   for node=0 to 3
4     if node is a leaf then
5       if value eq 'A' area=area+leafarea
6     else
7       call area(leafarea)
```

We have now given our program a name – area. When it is called, the value of the current leaf size is passed as a parameter called leafsize. The program then sets up a local variable, called leafarea, which it sets to be the new size of leaves – a quarter of the previous value. The program then visits each of the current four nodes in turn, looking for leaves labelled A. If it finds another branch node, it calls itself again passing it the value of leafarea. The newly called version will receive this value and set up its own local variable to store the size of leaves at the next level down – although this is also called leafarea, it will be kept quite separate from the first leafarea variable. To complete this algorithm, we also need a program which will set the initial value of leafarea at the root of the tree, and call area for the first time:

```
program quadtree_area
layersize=16
call area(layersize)
```

When a computer program calls itself in this way it is termed recursion. As we can see, in this case it has provided a very elegant way of traversing our quadtree structure.

FURTHER READING

As with vector algorithms, there is a good deal in common between the algorithms for raster GIS and for raster computer graphics. Indeed, the common ground is much broader in this case, because similar methods have application in remote sensing, image processing, machine vision, pattern recognition and 2D signal processing. There is even some commonality with computer modelling techniques such as genetic algorithms and cellular automata, which also use a grid of cells as their basic data structure. A good general text on image processing is Rosenfeld and Kak (1982). Of the Remote Sensing texts, Mather (1999) contains a certain amount on the operation of standard functions such as filtering and image rectification. The standard computer graphics text, (Foley *et al*., 1990) covers raster as well as vector operations.

In this chapter, some simple algorithms were used to demonstrate the idea that raster data in quadtree format could be used to answer queries directly, without first converting the data back to a simple, raster array. Some further simple examples are covered in the NCGIA Core Currucilum Unit 37 (http://www.geog.ubc.ca/courses/klink/gis.notes/ncgia/toc.html). The two books by Samet (1990a,b) cover a much wider range of algorithms, such as raster overlay, and illustrate how these too can be implemented using quadtrees.

An area where GIS and image processing come together is in capturing data from scanned maps using image processing methods. De Simone (1986) was one of the first to use automated image processing algorithms to identify vector features on scanned maps. More success has been achieved using semi-automated methods, which combine the image processing abilities of the human eye and brain with tools such as automatic line followers. Devereux and Mayo (1992) made some important developments in this area in the case of capturing vector data, and the approach has been extended to raster data capture by Wise (1995, 1999). Wise (1995) also discusses some of the programming issues which arise when handling raster data on a machine where memory is a limitation.

Something which has not been explicitly covered here is the conversion between vector and raster. The two articles by Peuquet (1981a,b) discuss some of the early thinking on this, when it was an issue which dominated a good deal of the GIS literature. Jones (1997), in Chapter 8, provides a summary of some of the main algorithms, including Bresenham's algorithm for rasterising vector data. This is a particularly important operation in computer graphics, since computer screens are raster devices, so drawing any type of vector linework requires a conversion to raster. Bresenham's algorithm is important because, like the example of handling Morton codes in the chapter, it takes advantage of the greater speed of integer arithmetic, especially when this involves powers of 2.

8 Spatial indexing

We saw in Chapter 5, that one of the key issues in the design of data structures and algorithms for spatial data is that they should be efficient. This means that even if the datasets are large queries can be answered in a reasonable time. This often leads to quite different solutions for vector and raster data because of their inherently different characteristics. However, one area where similar ideas can be applied to both vector and raster is in indexing objects by their spatial location, which is the subject of this chapter.

8.1 BINARY SEARCH TREE

The role of an index in computing is to speed up the process of locating one or more pieces of information. The idea is the same as with the index in a book such as this one. By looking up a topic in the index, it is possible to turn directly to the relevant page without having to read through the whole book. We saw in Section 7.1, how it is possible to extract the value for individual pixels in a raster layer by using the row and column number. In effect, the raster grid acts like a simple, locational index, allowing us to retrieve information according to position very quickly. However, things are not always as simple as this, and it is sometimes necessary to use additional data structures to speed up queries which need to retrieve objects according to their position.

In order to understand the indexing of spatial data, it is easiest to begin with a non-spatial example. For this we will return to the problem of searching a list of numbers, which was used to introduce the concept of algorithm efficiency in Chapter 5. Imagine we have the following 8 numbers shown in the upper half of Figure 8.1 and we wish to find out whether number 3 is one of them.

If the numbers are not in order, we are going to have to look at each one in turn. If it is not we will look at all 8 numbers. If it is, we will look at roughly half of them. In both cases, this is an $O(n)$ algorithm. However, if the numbers are sorted, we can perform a binary search. This was described

| 4 | 7 | 18 | 10 | 1 | 3 | 11 | 14 |

| 1 | 3 | 4 | 7 | 10 | 11 | 14 | 18 |

Figure 8.1 Operation of binary search. Upper table: 8 numbers in random order. Lower table: same numbers in numerical order.

in Chapter 5 but is worth summarizing again since the logic underlies much that is to follow. We begin by examining the middle number in the list. If it is what we are looking for, we have finished. If it is not the number we want, and it is greater than 3, we repeat the process on the first half of the list. If it is less, we examine the second half of the list. We can only divide the list into half three times, and we now have an $O(\log n)$ algorithm.

The pseudo-code for a procedure which implements this is as follows:

```
1   procedure BINARY_SEARCH(k)
2   Array SORTED[1..n]
3   first=1
4   last=n
5   found=false
6   repeat until first==last or found==true
7     middle=((last-(first-1))/2)+first
8     if(SORTED[middle]==k]
9       found=true
10    else if(SORTED[middle]<k)
11      first=middle+1
12    else
13      last=middle-1
14  if found==true or SORTED [first]==k
15    return middle
16  else
17    return -1
```

The procedure keeps two variables, first and last, to keep track of which part of the sorted array to deal with. Initially these point to the start and end of the array. Line 7 calculates the middle position, which in this case will be 5. The fifth element is 10, which is greater than 3. This causes the variable last to be set to the array element to the left of middle, so on the second time through the repeat loop, elements 1 to 4 will be searched. The middle in this case is number 3 which has a value of 4. Next time elements 1 to 3 will be searched and the middle one is the second element. This contains the search term and so the procedure will return a value of 2.

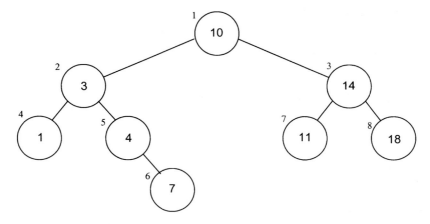

Figure 8.2 Binary search tree. Numbers inside circles are values held in nodes.
Numbers next to circles are node IDs.

If the repeat loop continues until first and last become equal, this means
that no further subdivision is possible and the repeat loop ends. A final
check is necessary to see whether this final element of the array contains the
value we are looking for. If not then a value of −1 is returned to indicate the
search item is not in the array.

We can also use a tree data structure to simplify this procedure and to
avoid the constant calculation of which element of SORTED to look at next.
These same 8 numbers are shown stored in what is called a binary search
tree in Figure 8.2. Like the quadtree, the tree is made up of two types of
nodes. Leaf nodes are those which have a number stored in them, but no
further numbers below them. Branch nodes also store a number, but in
addition they point to two further branches of the tree – one on the left and
one on the right. Every branch is organized so that all the nodes below it on
the left contain numbers which are smaller than it, and all the ones on the
right are larger.

Check for yourself that this is so in the case of Figure 8.2. The root node
contains the number 1. Every number down the left hand branch is less than
10. Every number down the right hand branch is greater than 10. The same is
true of the node containing the number 3 – this has 1 on the left and 5 on the
right. Figure 8.3 shows how this tree might actually be stored in computer
memory. Each node is actually a small array of 5 numbers as follows:

1   ID of the node
2   value stored in the node
3   pointer to the left child
4   pointer to the right child
5   pointer to the parent.

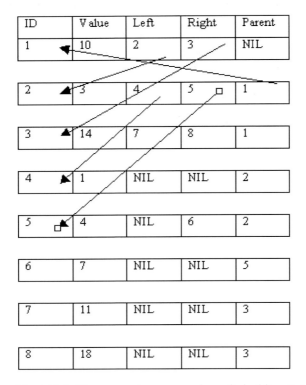

ID	Value	Left	Right	Parent
1	10	2	3	NIL
2	3	4	5	1
3	14	7	8	1
4	1	NIL	NIL	2
5	4	NIL	6	2
6	7	NIL	NIL	5
7	11	NIL	NIL	3
8	18	NIL	NIL	3

Figure 8.3 Binary search tree stored as a linked list. Pointers for nodes 1 and 2 are shown with arrows.

The IDs have been included as a simple way of illustrating the idea of pointers. Node 1 (the root) points to nodes 2 and 3. Since it is the root, it does not have a parent, so this pointer is stored as NIL. Node 2 has pointers to nodes 4 and 5 as its children and 1 as its parent. Node 4 is a leaf node, and so has the pointers to children set to NIL.

In practice, the pointers could be implemented in a number of ways. This tree could be stored in a one-dimensional array, as was done for the quadtree, in which case the pointers could be the number of the relevant array element. This is how the quadtree was stored in Section 6.2. In this case, there is no need for IDs and each node would be held in 4 units of storage. However, since each node has a pointer to its children and its parent, there is no need for them to be stored next to each other at all. As long as pointers contain the actual memory address of the child or parent node, it does not matter wherein memory this is. The advantage of this method of storing a tree is that it is not necessary to set up space in memory for the entire tree at the start of the program. If it is necessary to add extra nodes while the program is running, more memory can be allocated wherever it is free, and the address of this new memory put in the appropriate pointers.

Whichever method is used to store the tree in memory, the important thing is that it is a simple matter to search it to find a particular element. In outline, the procedure is as follows:

```
1  Look at value for current node
2  If same as search value STOP
3  If search_value<current SEARCH(left-child)
4  Else                    SEARCH(right-child)
```

In more detail:

```
 1  SEARCH_BST(T,k)
 2  if T==NIL then
 3    print 'Not found'
 4  else if k=value[T]
 5    print 'Found'
 6  else
 7    if k<value[T] then
 8      SEARCH_BST(left[T],k)
 9    else
10      SEARCH_BST(right[T],k)
11    return
```

The procedure passed a pointer T to the root of the tree. When the procedure is first called T should not have a value of NIL so line 4 will be executed. This tests whether the value at the current node matches the search item, k. If it does, the search has been successful. If not, SEARCH_BST is called again, this time with a pointer to the right or left sub-tree from the current node. Assume this procedure is called to see whether 3 is included in the list. The sequence of calls to SEARCH_BST would be:

```
1  SEARCH_BST(1,3)
2  3<11
3    SEARCH_BST(2,3)
4    3=3
5    Found
```

On line 5, the second call to SEARCH_BST succeeds. At this point, the procedure returns to the procedure which called it. This was actually the previous call of SEARCH_BST, which at the time was on line 8. This call of the procedure then returns to the program which called it, terminating the process.

The binary search tree shown in Figure 8.3 is well balanced, which means that the majority of branch nodes have two children. This is equivalent to dividing the original sorted list into half at each stage of the search, and so the search process will be O(log n) in this case. However, it depends on how the tree structure is built from the original data as to whether this is true. To take an extreme case, imagine taking the sorted list of eight numbers, and adding them to the tree one at a time in sorted order. Every node would have a single, right-hand child node, because each successive number added would be larger than the previous one. The list would therefore be 8 levels deep, and no better for searching than the original, brute force search through the sorted list. This is an extreme case of course, and unlikely to arise in practice. To avoid it, a number of modified forms of the binary search have been developed. Alternatively, it might be worth sorting the original list of numbers, which can be done on O(n log n) time, since this makes it easy to build a balanced tree. The root of the tree is the middle value in the sorted list, since by definition this has equal number of points on either side of it. If we apply the same logic to each half of the list at each subdivision, we will create a balanced tree.

This example shows that the key to efficient searching is to sort the items in order and use an algorithm which can make use of this order. Searching is an important task with spatial data as much as with any other sort of information. Typical queries include finding all objects which fall within a particular rectangular window, or finding the nearest object to a particular point. The problem is that because spatial data has at least 2D, there is not always a natural order which can be used to sort the location of objects. If objects are sorted in terms of their Y coordinates, then two objects which have very different X coordinates, and are thus very distant from each other, may end up very close to each other in the sort order. Indexing techniques for spatial data therefore have to deal with both the X and Y coordinates.

8.2   INDEXING DATA WITH A k-d TREE

The first mechanism we will consider for indexing vector data is based on the idea of extending the concept of the binary tree to more than one dimension, to create what is called a k-d tree. The k originally referred to the number of dimensions being handled, since the basic idea was not limited to 2D but could apply to three or even higher number of dimensions. In theory, then one could talk about a 2-d tree, but the term k-d tree has become generally accepted for referring to this data structure.

For simplicity, let us start by considering the one dimensional case. Figure 8.4 shows a series of points located along a line. The problem is to identify which of them lie within the range shown by the dotted lines. Clearly, the simple approach is to compare the X coordinate of each point with the X coordinates of the ends of the range. However, by utilizing a search tree, we can reduce the number of points for which we have to make this

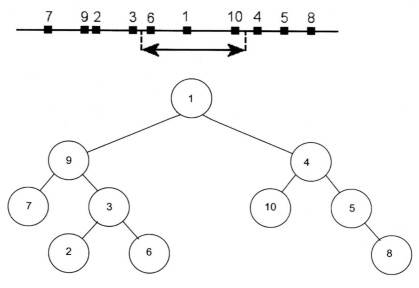

Figure 8.4 Ten points located along a single line. The binary search tree is based on the *X* coordinates of the points.

comparison. A standard binary search tree can be used to store the *X* coordinates of the points as shown in Figure 8.4. This has been constructed in exactly the same way as the example in the Section 8.1, but using the *X* coordinates of the points rather than their IDs. Hence all the points on the left hand side of the tree are to the left of the root, which is point 1.

To find out which points fall within the range, we start at the root, which is point 1, and test whether this is inside the range. It is, which means that points to both left and right may also be inside the range so we must examine both the nodes below the root. We begin with the left child which contains point 9. This is to the left of the left hand end of the range. This means we do not need to investigate the tree below the left hand child of 9, because none of the points in this tree can be to the right of point 9. We do have to examine the right hand child, which is point 3. This is also to the left of the range, so we only have to examine its right hand child – point 6. When we look at the right hand branch of the tree below the root, we find that since point 4 is to the right of the range, we only need to consider its left hand child, point 10 – points 5 and 8 can safely be ignored. The algorithm for this search is the same as a basic binary search, except that instead of looking for an exact match with the search key, we need to look for whether the *X* coordinate of the candidate point falls within the range. In this simple example, of the 10 points we only need to inspect five of them to discover the three which actually fall in the range.

The *k*-d tree simply extends the logic of the binary search to extra dimensions. So how does it manage to combine the *X* and *Y* coordinates into a single sort order? The answer is that it doesn't. It constructs the tree by considering the *X* and *Y* coordinates alternately. In Figure 8.5, the same 10 points have been given a *Y* coordinate. The root of the tree is simply the first point in the sequence – point 1. Point 2 is added next. We compare its *X* coordinate with the *X* coordinate of the root – since it is less, it becomes the left hand child of point 1. Point 3 is next and again we start at the root of the tree. Point 3 also has a smaller *X* coordinate than the root so will need to go in the left hand half of the tree. There is already a point in the node immediately below the root, so we look to see if this has any children. It does not, so point 3 will be added as a child of point 2. The children of the root are assigned according to their *X* coordinates. We are now one level down from the root, so to determine whether 3 is the left or right hand child of 2 we compare their *Y* coordinates, rather than their *X* coordinates. The rule is that the smaller value goes in the left hand half of the tree. Point 3 is above 2, which means it has a larger *Y* coordinate, and will therefore be the right hand child.

The same procedure is followed for all the other points. To insert a point, we start at the node of the tree and proceed down until we reach the first available position which is not already occupied. At the root level, we decide which branch to look down by considering the *X* coordinate of the new point and the point currently in the tree. After this, at each level down the tree we alternate between considering the *Y* and *X* coordinates.

The range query for the *k*-d tree is then essentially the same as the procedure for searching a binary search tree:

```
1   procedure range_search (T,RANGE,level)
2   if T==NIL return
3   if T inside RANGE Mark T
4   if level is even then
5     if(T[x]>RANGE[x1]) then
6       call range_search(T[left],RANGE,odd)
7     if(T[x]<RANGE[x2]) then
8       call range_search(T[right],RANGE,odd)
9   else
10        if(T[y]>RANGE[y1]) then
11          call range_search(T[left],RANGE,even)
12        if(T[y]<RANGE[y2]) then
13          call range_search(T[right],RANGE,even)
14    return
```

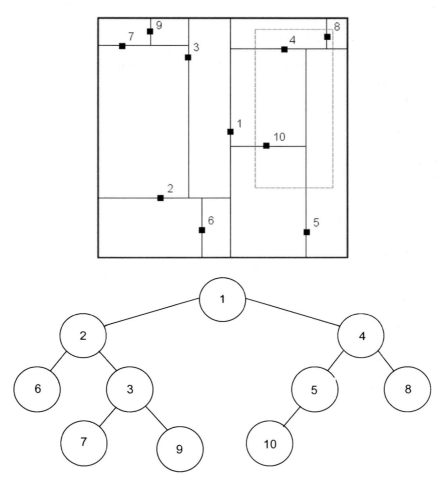

Figure 8.5 *k*-d tree. Above – map of points. The search window for the range search example is shown with a dashed line. Below – tree structure after all 10 points have been inserted.

Now that we have a 2D search the range is a rectangle, defined by the X and Y coordinates of its lower left and upper right corners as shown in Figure 8.5. When range_search is first called in this case, it is passed a pointer to the first node in the tree which is point 1. This is not inside the range, so is not marked. On this first call, the variable called level is set to the value even. The X value of the current root is then compared with the

X values of the range window, as was done for the one-dimensional case. In the case of the tree in Figure 8.5, the test on line 4 will fail, which means that none of the branches to the left of point 1 will be tested. On line 6, point 1 is to the left of x2, so `range_search` is called to perform a search starting with the right hand child of point 1. Note that this time the level variable will be passed as 'odd', so that it is the *Y* coordinates which will be considered when looking at point 4.

When point 4 is checked it will be found to be in the range window, so it is marked. Since it is in the window, both its left and right children will have to be examined. When point 5 is checked, it is not in the window. However, because the *X* value of point 5 is within the range of *X* values of the window, both its children will be searched. Point 5 does not have a right hand child, so this call to `range_search` will terminate immediately. However, point 10 will be found to be within the range and will be marked.

As with the one dimensional case, the use of the tree structure, sorted according to the *x* and *y* coordinates, has reduced the number of points which need to be searched. In this example, of the 10 points only 5 have been considered, of which 3 are in the window. A brute force algorithm for range searching would compare the *x* and *y* coordinates of each point with the coordinates of the window and would be an O(*n*) operation.

The balanced *k*-d tree has a depth of O($\log n$) as with many binary structures. However, this is not the efficiency of the search algorithm. Consider two extreme cases. If the search window happened to contain all the points, then the entire tree would have to be searched, so `search_range` would be called *n* times. The other extreme is represented when the search rectangle contains none of the points. In this case, `range_search` would only ever have to consider one branch of the tree at each level, so would run $\log n$ times at most. Clearly, therefore the efficiency depends both upon the number of points in the tree, and on the number which fall in the window. De Berg *et al.* (1997) show that the efficicency is actually O ($\sqrt{n} + k$) where *k* is the number of points falling inside the window.

8.3   INDEXING VECTOR DATA USING A QUADTREE

Since the quadtree is based on the idea of subdividing 2D space, can we also use this data structure to help with spatial queries? The answer is that we can, and to illustrate how let us consider a simple example. Figure 8.6 shows part of an imaginary spatial database for a nature reserve. The features shown are an information centre, a footpath which crosses the area, three lakes, and a series of hides for use by birdwatchers. Notice that in this case,

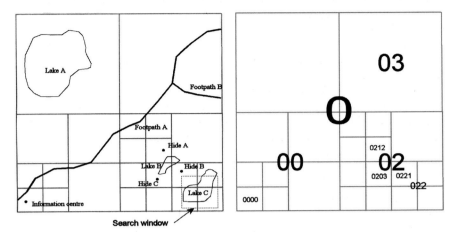

Figure 8.6 Left: map of nature reserve showing quadtree leaves (grey solid lines) and the search window (grey dashed line). Right: Morton addresses of leaves containing features.

we have examples of points, lines and areas to deal with, and not just points as in the previous example.

The data we are dealing with is vector and not raster, so we cannot use the sort of quadtree which was described in Chapter 6 to store the locational data. However, we can use the ideas underlying the quadtree to create a spatial index, which we can use to help us retrieve objects from this database based on their location. The key is to identify the smallest quadtree node which completely encloses each of the objects in our database, as shown diagrammatically in Figure 8.6. Consider the lake in the north west corner of the area (Lake A). If we subdivide the whole area into four quadrants, then quadrant 01 will contain this lake. A further subdivision of this quadrant would intersect with the lake, and so this is the smallest quadtree leaf which will enclose it. The Morton address for this quadrant is 01, and we can use this as the spatial index for this lake. The address itself tells us two things about lake A. First, we know which part of the map it is in, since it falls in the north west quadrant at the first level of subdivision. Second, we know roughly how large it is, because it cannot be any bigger than a quarter of the total area of the map. This is only approximate information of course. For instance lake B, in the south-east section of the map, also needs a quadtree leaf of the same size since it happens to lie in the middle of leaf 02. However, the Morton code can be far more useful than providing approximate information. Figure 8.6 shows the quadtree

leaf which encloses each object, and the corresponding Morton codes
are shown in the table:

Feature	Morton address of smallest enclosing quadtree leaf
Footpath A	0
Lake A	01
Lake B	02
Footpath B	03
Lake C	022
Hide A	0212
Hide B	0203
Hide C	0221
Information centre	0000

Assume we wish to find out which objects fall within a rectangular window
such as the one shown by a dotted line in Figure 8.6. This is the same as
the range query we saw in Chapter 7, but applied to objects other than
points. We can use our quadtree addressing to identify the quadtree leafs
which will contain objects that might fall within the query window. The
query window itself is contained within quadtree leaf 022 and anything else
which is wholly contained within this leaf might fall within the query
window. Lake C has exactly this Morton address, and so will need to be
considered. Note that as with the MER test, this simply identifies the fact
that Lake B may fall within the window – a more detailed intersection test
will have to be carried out to see whether it does or not.

Figure 8.6 shows that hide C is also contained within this quadtree leaf.
We can tell this from its Morton address of 0221 which makes it clear that
this is one of the child nodes of 022. It should be clear from this that, that
objects in any of the other child nodes of 022 would also be possible
condidates for inclusion in the window. We also need to consider objects
which are larger than the query window, since part of them may fall inside
the window. However, in doing this we only need to consider quadtree
leaves which are directly above 022 in the quadtree structure – its immediate parent, leaf 02, and the root itself which is leaf 0. This means that lake B
(address 02) and the footpath A (address 0) would have to be checked for
intersection. However, none of the other objects would need to be checked,
which means that of the eight features in the simple map, only four of them
would have to be considered.

The sort of data structure which was described in Chapter 6 for storing
raster data in a quadtree will not really be suitable for this situation for two
reasons. In a raster layer, the variable which is being stored has a value
everywhere, and only has one value at each point in space. Neither of these
things is true with vector data. Figure 8.6 shows that some of the quadtree
leaves do not contain anything, while others contain more than one object.

The basic tree structure would therefore have two weaknesses. Some of the nodes in the tree would not contain anything, while others would have to contain a variety of information. For example, the root node itself would have to have pointers to its four child nodes, plus information on the fact that this is the smallest node which encloses the stream.

In fact we do not need to use the quadtree data structure at all since our query can be handled entirely using the Morton addresses themselves. The algorithm to find the objects which might intersect with node 022 is to find:

```
1   All objects with Morton address starting 022.
2   All object with Morton address 02.
3   All objects with Morton address 0.
```

Let us consider the first condition. This is a very easy query to answer, if we consider the Morton addresses as a set of four digit numbers. To do this, we extend all the addresses to four digits by adding extra zeroes on the right hand side. If we sort these numbers into numerical order, we get the list as shown in the following table.

Feature	Four digit Morton address of smallest enclosing quadtree leaf
Footpath A	0000
Information centre	0000
Lake A	0100
Lake B	0200
Hide B	0203
Hide A	0212
Lake C	0220
Hide C	0221
Footpath B	0300

The addresses we are interested in are those which start with 022. When we get to an address which is 0230 or larger, we have found all the addresses we need. We have already seen a data structure which is ideally suited to handling sorted numbers like this – the binary search tree. If we store our Morton addresses in a binary search tree, we will be able to use the SEARCH_BST procedure described in Section 8.1 to find address 0220. All the addresses which are larger than this are stored in the right hand part of the tree below this point. Whenever we reach an address which is 0230 or larger, we can stop searching that part of the tree.

Converting the Morton addresses to four digits has made it possible to sort them into order and use an efficient method to find particular addresses. However, it has made it more difficult to find the leaves which lie above 022 in the quadtree structure because in order to do this, we need to know what level each leaf is in the quadtree. When the Morton addresses were stored as variable length numbers, as in the first table, the number of digits in the address told us how large the leaf was – hence leaf 02 is twice as large as leaf 022. Once we convert the addresses to four digits, we lose this information, so that for example the information centre and Footpath A both now have a Morton address of 0000, even though one is a point and one crosses the entire area.

There are two solutions to this problem. One is to use digits 1 to 4 to identify the four quadrants rather than 0, which leaves zero free to be used as a filling digit. In this case, Footpath A would have address 1000, while the information centre would be 1111. The alternative is to store the level for each Morton address explicitly, starting at 0 for the root level, as shown in the table below:

Feature	Four digit Morton address of smallest enclosing quadtree leaf	Level
Footpath A	0000	0
Information centre	0000	3
Lake A	0100	1
Lake B	0200	1
Hide B	0203	3
Hide A	0212	3
Lake C	0220	2
Hide C	0221	3
Footpath B	0300	1

Once we have the level, our second query changes from being

```
All object with Morton address 02.
```

to

```
All objects at Level 1 with Morton address 0200
```

The address we are looking for has been changed to the four-digit format, and the query changed to check for the level. Our original search address is 0220 . We need some way of extracting the first two digits of this address, to create the value 0200 which will be used in this search procedure. In Chapter 7, we saw how it was possible to extract individual digits from a Morton address by simple manipulations on the word holding the address. Since the digits of the Morton address are all in base 4, dividing the number which holds the address by 4 shifts all the digits to the right one position and loses the right-most digit. If we then multiply by 4, this will move the digits back to their original position, but the right-most digit will now be a zero.

To divide by 4 all we need to do is shift all the bits in the word two places to the right. To muliply by 4 we shift all the bits two places left.

Our algorithm for finding the Morton address of a feature at any level is therefore

```
Shift address right 2x(maxlevel-level) bits
Shift address left 2x(maxlevel-level) bits
```

Variable `maxlevel` is the maximum level in the quadtree. In our case, this is three – there are four levels of subdivision, numbered from 0 to 3. Let us see how this would work for our query.

Address 0220 will be stored in binary as follows:

```
00101000
```

This is the address at level 3. To get the level 1 address, we shift the word right by $2 \times (3 - 1) = 4$ bits and then left again by 4 bits. The first shift produces:

```
00000010
```

and the second shift results in:

```
00100000
```

This is address 0200 which is 0220 with the two right hand digits set to 0. We can then search for this using our binary search tree, and will find that we have a record representing Lake B. The same sort of procedure could be used to extract the address 0 which is the third part of our original search query. However, this is not really necessary since we will always have to check features whose smallest enclosing leaf is the entire map.

We have now seen that the ideas underlying the quadtree can be used to provide efficient spatial indexes for vector data. This chapter can be concluded by looking at one further application of Morton order to raster data.

8.4   INDEXING RASTER DATA USING MORTON ORDER

In Chapter 7, the quadtree was presented as one means of saving space in the storage of raster data. This feature of quadtrees is most important when we have raster data which contains large, uniform areas. A good example is a rasterized version of a thematic map, such as a soil or landuse map, in which it is not uncommon to find large areas with the same soil type or landuse category. However, not all raster data layers possess this characteristic. For instance, in a gridded DEM, in which the value in each raster cell is elevation, neighbouring pixels are unlikely to have exactly the same value. However, there are two ways in which the ideas underlying the quadtree can be used with datasets like this. First, the different levels of the quadtree can be used to represent the same set of data at different levels of detail. Second, the pixel addressing system implied by the use of Morton addresses can be used to increase the efficiency of accessing large files. We will examine each of these in turn.

The fundamental idea behind the quadtree is that where there is no spatial variation in the values in neighbouring pixels, these pixels can all be represented by one large pixel. However, there are situations in which we do not actually need to see all the details in a raster layer, such as if we want a broad overview of the whole layer, or if the layer is simply too large for every pixel to be drawn on the screen. This means we need to simplify the image, which in turn might make it suitable for storage in a quadtree.

A variety of methods can be used to produce a simplified version of a series of pixels. Assume that we have an image and we wish to simplify it so that it is only a quarter of its original size in both X and Y. This means that every block of 16 pixels in the original image will have to be represented by just one pixel in the simplified image. The simplest method is to take one pixel from each block of 16 and ignore the rest. Like all simple algorithms, this will not necessarily produce good results. However, it will be quick. If all that is needed is some way of providing a broad overview of an image to allow a user to be able to zoom in on the area of interest, this method may suffice. It also has the advantage that it is not necessary to pre-process the image. A slightly more sophisticated technique will represent each 16 pixels by some average of their values – with numerical data this might be the arithmetic mean, while with integer data the median may be more appropriate. Many computer graphics packages use what is called bicubic spline, which models the pattern of variation within the block using a mathematical surface, and is better at preserving important characteristics of the original image, such as edges and highlights. Even more sophisticated methods are used in image compression standards, such as JPEG (Joint Photographic Experts Group), and in commercial products such as MrSID (Multiresolution Seamless Image Database).

The advantage of storing these simplified versions of the original image in a quadtree is that if a user decides to zoom in on a particular area for more detail, the quadtree structure provides a simple way to identify which pixels need to be retrieved in their full original detail. Thus, the quadtree structure can be used to store the same raster data in several levels of detail, from the broadest overview down to the finest detail available.

The final topic which will be considered in this chapter brings us back to one of the key reasons that issues like spatial indexes are so important – the increasing need for GIS to handle large data files. The array data structure for raster data has a great advantage that it is a very efficient means of accessing the values for any pixel in the array. Since retrieving a value is simply a matter of working out the address of the pixel in memory, it is equally fast no matter how large the array. Well, not quite. It is equally fast as long as the whole array can be held in memory. If this is not the case, we can try and reduce the size of the image, using data structures such as run length encoding and the quadtree. However, raster layers which do not have areas of uniform value will not be reduced in size by this method.

Therefore, the alternative is to read part of the layer into memory from secondary storage, process it and write it back to secondary storage. This is then repeated with the next part of the image. If we use this approach, we are not limited to dealing with the array in its normal row order, in which the whole of one row is stored, followed by the whole of the next row. For example, imagine if we had a query which required us to deal with only the first few columns, but for every single row in the image. If we handled the image in row order, we would have to read the whole of every row into memory, most of which would not be needed. This is the same problem we have seen before with indexing spatial data. By organizing the data, a row at a time, we ensure that pixels with similar Y coordinates are likely to be stored near each other, but that pixels with similar X coordinates may be stored a long way from each other in the file. An alternative therefore is to store the pixels in a sequence which combines the X and Y positions of their location and one possibility is to use the Morton address. If we calculate the Morton address of each pixel in a raster layer, and then store the pixels in the order of their Morton address, the pattern in which pixels will be stored will be as shown in Figure 8.7.

This shows that instead of being stored row by row, pixels are stored block by block. There are still cases where pixels which follow each other in the sequence are distant from each other in space, but in general pixels

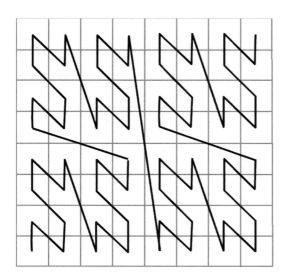

Figure 8.7 Sequence of pixels when stored in order of Morton address.

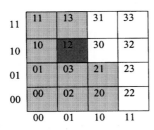

Figure 8.8 Raster layer showing Morton addresses of pixels. See text for explanation of shading.

which are close in the sequence tend to be closer in space than is the case for row ordering.

The sequence in which pixels are stored is called a scan order. The key to using any scan order is that it should still be possible to determine the spatial relationship between the pixels. For instance, the first pixel in the sequence in Figure 8.7 will have a Morton address of 000. But what will be the address of the pixel immediately to the east? Or two positions to the east? Unless we can answer these questions, the scan order is of little use.

To see how this works with Morton order, let us assume we want to compare a pixel with its immediate neighbours. One possible application of this is for the calculation of a slope angle from a DEM, which is described in Section 10.1. For simplicity, we will assume we are dealing with the very small raster layer shown in Figure 8.8 – just 16 pixels. The numbers along the edges are the X and Y coordinates of the columns and rows in binary. The numbers inside the pixels are the Morton addresses in base 4. The pixel we are interested in is in column 1, row 2 and has been shaded a dark grey in Figure 8.8.

We will also assume that we have an extremely small computer, which can only store 10 pixels at a time in memory! Therefore, as well as needing the address of the neighbours, we need a way of determining whether they are currently in memory. The first 10 pixels in Morton order have been shaded in the diagram. Since the pixels are read in Morton order, we know that any pixel with a Morton address between 00 and 21 will be in memory – all pixels with addresses larger than this will be on disk. The key to finding out whether the neighbours of pixel 12 are in memory is therefore to calculate their Morton addresses. Let us take the example of the pixel immediately to the east of 12. To calculate the Morton address of 12, we took the X and Y coordinates of this pixel in binary and interleaved their bits:

$$X = 01$$

$$Y = 10$$

$$\text{Morton} = 0110_2 = 12_4$$

The coordinates of the pixel immediately to the right will be $X + 1$, Y. Therefore, we can calculate its Morton address:

$$X + 1 = 01 + 1 = 10$$

$$Y = 10$$

$$\text{Morton} = 1100_2 = 30_4$$

Using the same aproach we can calculate the Morton address of the other seven neighbours. In fact, if we needed access to all eight we would start by looking to see if the south-west neighbour was in memory. If it is not, we know that none of the neighbours are, and we will have to read them in. By examining the largest Morton address currently in memory we can work out how many more pixels we need. If the south west neighbour is in memory, we would check the north-east neighbour. If this is also in memory, then all eight must be and we can proceed. If not, we will have to read the necessary pixels from memory.

This subject of spatial indexing is a very important one for practical GIS applications, especially as systems becoming more widely used in major organizations, and the size of the databases which need to be processed increase. This chapter has only scratched the surface of this subject but hopefully it has conveyed a flavour of the special problems posed by the two dimensional nature of spatial data.

FURTHER READING

Both Worboys (1995) and Jones (1997) provide good descriptions of the topics covered in this chapter, including some techniques not described here. Both versions of the NCGIA Core Curriculum have units on spatial indexes (http://www.geog.ubc. ca/courses/klink/gis.notes/ncgia/toc.html., http://www.ncgia.ucsb.edu/pubs/core.html). For a recent overview of the entire field, see van Oosteroom's chapter in the second edition of the Big Book (van Oosteroom 1999).

Cormen *et al.* (1990) is a good place to start for a description of various tree structures for non-spatial data, including binary search trees. De Berg *et al.* (1997) describe the use of *k*-d trees for range searching, and also methods for searching in non-rectangular windows. As with anything related to quadtrees, the two books by Samet (1990a,b) give very thorough coverage. Samet (1990a) gives more details on finding which leaves in a quadtree are neighbours, for the more general case when the leaves are not all the same size. Samet (1990b) describes the point quadtree, which is similar to the *k*-d tree and its application to range searching. The quadtree data structure described in Section 8.3 is based on the structure used by Shaffer *et al.* (1990) for a GIS called QUILT.

Abel and Mark (1990) and Goodchild and Grandfield (1983) discuss the relative merits of a series of different scan orders for a range of queries and Bugnion *et al.* (1997) provide a general review of scan orders. Dutton (1999) has worked on the use of a triangular mesh for providing a uniform locational code for any point on the Earth's surface. There is also a web page (http://www.spatial-effects.com/SE-Home1.html) which illustrates these ideas (along with a variety of other things spatial). Image compression is a good example of the way that general techniques for handling raster imagery can be applied to raster GIS data. Kidner *et al.* (2000) describe a TIN-based method for providing multi-resolution access to DTM data. Large geographical datasets are increasingly being made available in a format called MrSID which had its origins in the development of a system for storing and scanning images of fingerprints for the FBI. The format makes use of sophisticated image compression algorithms, and a hierarchical structure which allows browsing at different levels of detail. Several examples are available on the Lizard Tech web site (http://www.lizardtech.com/).

9 Data structures for surfaces

Chapters 1–8 of this book have introduced the main issues involved in the handling of vector and raster data on the computer. Many of the main data structures for handling these two forms of spatial have been described as well as a selection of algorithms for carrying out some fundamental operations. In chapters 9, 10 and 11, we will consider two types of spatial data which can be represented using either vector or raster data. In the case of surfaces, we will see that raster data structures and algorithms have some advantages over vector, and in the case of networks the opposite is true.

So far in this book, we have concentrated solely on data structures and algorithms suitable for 2D data. However, the third dimension is extremely important in many types of spatial data, and handling this extra dimension has required the development of specific data structures and algorithms. In the real world of course, objects exist in 3D space but for many purposes it is possible to ignore the third dimension, and produce a model which is still a useful approximation of reality. Maps in general are a good example of 2D models of 3D space, but topographic maps also show one aspect of the third dimension – the elevation of different points. On most modern map series the representation is in the form of contours, which as well as allowing an estimation of the height of points also portray general characteristics of the surface such as the steepness and direction of slopes.

Contours are not a true 3D representation. They can only be used to represent the upper surface of the landscape, and cannot show the existence of overhangs, or the shape of true 3D objects, such as geological strata. However, the surface itself plays an extremely important role in many natural and human processes. It is a key control in the distribution of both heat energy and water across the landscape, and thus is a key factor in the spatial pattern of climate, soils and vegetation. It is also a key control on the distribution of human activities, and an important factor in the location of settlements, businesses and transport routes. For these reasons data structures have been developed for representing the land surface in a GIS, and algorithms developed for displaying and analysing them.

The key characteristic of the land surface is that the main attribute – elevation – varies continuously across space. This is also true of a number of other spatial phenomena, such as climatic parameters (e.g. temperature, rainfall) and soil characteristics (e.g. depth, acidity), and the same data models are also used to represent these too. However, for simplicity, it will be assumed from now on that we are dealing with the land surface.

9.1 DATA MODELS FOR SURFACES

That fact that we are trying to represent a surface has two important consequences. First, since elevation varies continuously it is not possible to measure it everywhere. All measurements of elevation, no matter how closely spaced, are samples. Second, what is of interest are not the properties of the sample points themselves, but of the surface, and this means we have to form a model of the surface in the computer. In storing this sample data inside the GIS it is not enough to simply record the location and value as might be the case if the values related to discrete objects such as lampposts or houses. The values must be stored in such a way that it is possible to use them to derive useful information about the properties of the surface, such as the height, slope and aspect at any point. To see what this means consider, Figures 9.1 and 9.2. In Figure 9.1, the points represent the location of customers for a store, whereas in Figure 9.2, they are samples of height from a ridge, shown schematically using contours.

With the customer database, most queries can treat each point individually. For instance, to find out how many customers live within a 5 km radius of the store, the X and Y coordinates of the store can be used with the X and Y coordinates of each customer in turn to calculate the crow fly distance between the two. This can be done for each point in turn, and all we need to know is the location of the customers and of the store.

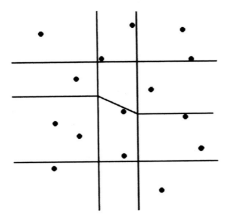

Figure 9.1 Points representing customer locations. Lines represent roads.

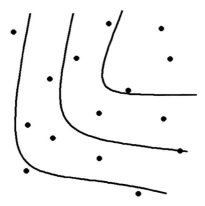

Figure 9.2 Points representing elevations on a ridge. Lines represent contours.

Compare this with what appears to be a very similar query using the point samples from the surface in Figure 9.2 – finding out the area of land which lies above 1000 m (or to be more precise – the area as projected onto a horizontal surface, as opposed to the true surface area). If we simply select all points with elevations above 1000, this will not answer the query, for two reasons. First, in order to estimate the area, we need to be able to work out how much land lies between the sample points we have selected. This means that we need to know where the points are in relation to one another, so that we can join them up. If we do this, then as Figure 9.3 shows, we hit the second problem, which is that our sample points do not define the edge of the area we are interested in. What we actually need is to calculate the area above the 1000 m contour, not the area between the sample points we have which happen to be above 1000 m. This means we need to use our

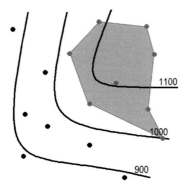

Figure 9.3 Problems with answering surface queries using point data.

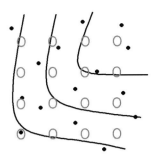

Figure 9.4 Grid data model for surface data represented by grey, open circles. The lines are contours and the dark dots are spot heights either of which can form the source date for creating the grid.

sample points to estimate where the 1000 m contour runs, and then use this line to answer the query.

In other words, as well as storing the location and elevation of each point, it is also necessary to store information about the relationship between the points. There are three main models which have been developed which can store this additional information. The commonest is the grid model, in which the original sample heights are used to estimate the height at a series of points on a regular grid as shown in Figure 9.4.

In the case of the grid model, the relationship between the points is very straightforward. Every point has eight immediate neighbours (except for points along the edges of the grid) and the distances between them are either the grid resolution, or in the case of diagonal neighbours, this value times the square root of two. This means that there is no need to store this information explicitly, and the data can be stored as a standard raster grid. Using the grid model we can provide an approximate answer to the query about the area of land above 1000 m simply by selecting all grid points with a value above 1000 m. Since the points are on a regular grid we can assume that the height values represent the 'average' height for the square surrounding each point, so we simply multiply the number of points selected by the square of the grid spacing. Assuming the grid points are relatively closely spaced, this will give an answer which is approximately correct. A more accurate answer could be obtained by interpolating the 1000 m contours line between the grid points.

The second model is the Triangulated Irregular Network (TIN). This takes the original sample points and connects them into a series of triangles, as shown in Figure 9.5 (the way in which the points are connected into triangles is described in Section 9.3). The relationship between the points is now stored using this triangular mesh.

In order to answer the query about land above 1000 m, one possibility would be to select triangles with at least one corner above 1000 m. For those

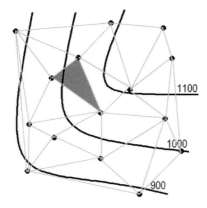

Figure 9.5 TIN model for surface data. Triangles, represented by grey lines, have been created by joining the spot heights.

with all three corners above 1000 m, the area of the triangle can be calculated very simply. Where a triangle has one or more corners below 1000 m, as with the shaded triangle in Figure 9.5 then some estimate would have to be made of how much of the triangle's area fell below 1000 m. This triangle has two corners below 1000 m. Inspecting each side of the triangle in turn it is easy to discover which sides cross the 1000 m contour and since the height at each end is known, it is possible to estimate how far along the side the 1000 m contour line will cross. The crossing points on the two edges can then be joined to produce a quadrilateral, whose area can be calculated. This will give reasonable results as long as the triangles can be assumed to have flat surfaces – where the terrain is highly curved, or where the triangles are large, a more sophisticated approach would be needed. However, what this simple example illustrates is the way in which the connection between the points is the key to answering the query.

The final model takes contours as its starting point. Contours are not only useful as a graphical means of displaying the nature of the surface, they are also equipotential lines for the potential energy possessed by water flowing across the landscape. Since water flows down the steepest slope, it will always cross a contour line at 90°. Therefore if a series of lines is drawn down the slopes, normal to the contours, then these, plus the original contours, define a series of 'patches' which define the surface as shown in Figure 9.6. (Note: for simplicity these lines are shown as approximately straight – in fact, they will be curves, unless the slope is completely planar between the two contours).

This is the most complex of the data models for surfaces, because the basic elements used to define the surface are no longer of uniform shape as they were in the case of the grid model and TIN. Information has to be stored, not only about the height at the corners of the patches, but about the

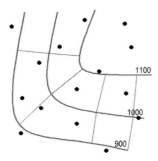

Figure 9.6 Contour-based model for surface data.

shape of the edges and how the height varies across the patch. The models to do this have been developed for representing solid objects in Computer Aided Design (CAD) software, since the method is well suited to represent smoothly curved objects such as car bodies or engineering parts. However, it is not available in standard GIS software packages, and is only used by a small number of researchers, so it will not be considered any further in this book. However, it is the model which would most easily give an accurate estimate of the area of land above 1000 m, since this is simply a matter of selecting all patches where both bounding contours have a height of at least 1000 m, and summing their areas.

This brief summary has indicated that there are two distinct sets of algorithms in the case of surface models in GIS. The first are the methods used to construct the model from the original data – to estimate the height at the grid points for the gridded model, and to construct the triangulation for the TIN model. The second are the methods used to answer queries using the constructed surfaces. A key issue in both cases is that of error. Since the surface can never be sampled or represented completely, the results of any analysis will always be estimates and will contain some element of estimation error. Keeping this to a minimum is one of the key issues in the design of algorithms for handling surfaces.

Finally, a note on terminology. There is a good deal of confusion in the GIS literature about the terminology for models of surfaces. All three models described here are examples of a DEM – a model of elevation in digital format. Many authors use the term Digital Terrain Model (DTM) as being synonymous with DEM, while others suggest that a DTM is distinct from a DEM, in that it models other features of the surface in addition to its height, such as its curvature. According to this definition the contour-based model is a DTM, the TIN is a DTM if the triangle's edges are placed in such a way that they represent breaks in the surface and the gridded model is a DEM. To further confuse matters, both DEM and DTM are often used without qualification to refer solely to the gridded model, since this is by far the commonest model in use. In this book, it will always be clear from the context whether it is the gridded DEM or the TIN which is being referred to.

9.2   ALGORITHMS FOR CREATING GRID SURFACE MODELS

In creating a gridded DEM, the task is to use the sampled elevation data (such as the points or the contours shown in Figure 9.7) to estimate the elevation at a series of points spaced on a regular lattice, as shown in Figure 9.7. The algorithm only needs to be able to interpolate height at a single point at a time, since we can just apply it repeatedly to fill the grid.

In thinking about how to design a point interpolation algorithm, it is instructive to think about how we might do the same task ourselves. First imagine that we have the contours shown in Figure 9.7, and we need to estimate the height at point B. This point lies between the 1000 and 1100 m contours, so its height will be somewhere between the two. Looking at these two contours, we can see that the point is about a 1/4 of the way between them, and is nearer the 1000 m contour, so we might estimate the height at 1025 m. Point C is obviously just over 1100 m. Since the slope has a fairly even slope, judging by the equally spaced contours, we could probably estimate that it will be around 10 m above the 1100 m contour giving a value of 1110 m. Point D is slightly more tricky, since it is on the ridge. It cannot be a full 100 m above 1100, otherwise there would be another contour, and given that slopes often have flattish tops, it may only be another 10–20 m above point C, so let us estimate it at 1115 m. This simple example illustrates two important points about this process. First, in

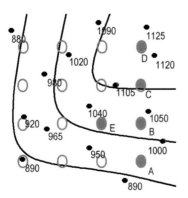

Figure 9.7 Small ridge showing sampled elevations (dark spots), contours and DEM grid points (grey circles). The shaded circles are referred to in the text.

estimating the height at an unknown point, we use information from nearby rather than far away. Second, in doing the interpolation we also use information about the shape of the surface in addition to the height data. Almost all computer algorithms for point interpolation use the first of these strategies, but differ in the way that they decide which points to use. However, they all do not use the second strategy, and this can lead to some major differences in the results they produce. As an example, let us consider one of the commonest algorithms for point interpolation, commonly called inverse distance weighting.

Consider the sample points and the DEM grid point labelled B in Figure 9.7. What value should we assign to the grid point? It seems clear that the value is likely to be similar to those for the sample points surrounding it, so one way of estimating the value is simply to take an average of the nearby data points. The selection of which points to use is an important issue which is discussed in detail below. To illustrate the principle of the method, let us use the four nearest points:

$$z = \frac{1050 + 1000 + 1040 + 950}{4} = 1010.0$$

This value seems too low, given how close the point is to the known value of 1050. We would also get the same result if we estimated a value for point E, since this is surrounded by the same four sample points. What we have to do is to give greater weight to the sample points which are near our grid point than those which are far away. This means that we need to measure the distance between the grid point and each of the sample points we use, and then use this distance to determine how important that sample points value will be. There are several functions we can use to do this, and two of the commonest are shown in Figure 9.8.

The equations for these functions are as follows:

Exponential

$$w = e^{-d}$$

Power

$$w = d^{-1}$$

In both cases d is the distance, and w is the weight and the functions produce weights which decrease towards zero as the distance increases. To apply this to the example above, we would calculate the distance to each point in turn, use this to calculate a weight using one of the formulas,

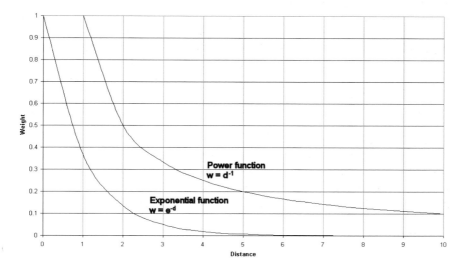

Figure 9.8 Two functions for assigning weights in point interpolation.

and then calculate a weighted average. For example, applying the exponential function to the previous example:

$$z = \frac{0.7 \times 1050 + 0.2 \times 1000 + 0.01 \times 1040 + 0.001 \times 950}{0.911} = 1038.80$$

Notice that we no longer divide by the number of points (4) but by the sum of the weights. We now have an answer which seems more reasonable, although it is possibly a little high. We could alter the result by changing the weighting function, so that the 1050 value does not have so much weight as the distant, smaller values. This can be done by including an extra parameter in the formula for the weights:

Exponential

$$w = e^{-p \cdot d}$$

Figure 9.9 shows that as p becomes smaller, the shape of the weighting curve changes so that distant points are given relatively greater importance. The difficulty is how to choose what value p should take. This is not the only problem. We also need to choose what sort of weighting function to use in the first place – two are shown in Figure 9.8, but which would be more appropriate for a given set of data? Both curves produce weights which decline asymptotically towards zero as d increases. (This means that they get closer to zero, but never actually equal zero until d equals infinity.) However, they behave very differently for small values of d. When d is zero, the exponential

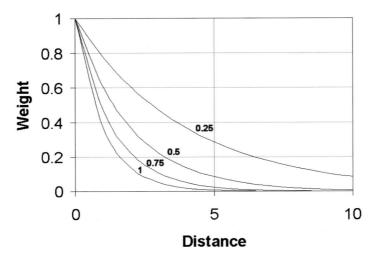

Figure 9.9 The effect of varying the exponent in the exponential decay curve.

function produces a weight of 1, but the power function returns a value of infinity. In practice, when d is zero, this means the point to be estimated is exactly coincident with one of the sample points, and so the z value of this is used as the interpolated value. However, for small values of d the power function will produce much larger weights than the exponential function, meaning that it assigns much greater importance to nearby points. However, it may be difficult for the GIS user to know if this is desirable or not.

There is also a further problem. In the simple example above, the nearest four points were used to make the calculations, to illustrate the principle of the method, but in practice the number of points to use would also have to be chosen. Most packages offer a choice of taking either a fixed number of points, or all points within a fixed distance, with default settings in both cases. The problem is how to decide which option is best and which number or distance to choose.

One solution to some of these difficulties is to use a method of interpolation called kriging, which provides answers to these questions by analysing the data itself. The first stage in kriging is to construct what is called a variogram, which is a graph which shows how similar the data values are for points at different distances apart. Points which are close together will tend to have similar height values, but this similarity will decrease as we pick points which are further apart. Eventually we will reach a distance at which there appears to be little relationship between the data values. This distance can then be used as a cut off point in selecting points to consider in the weighting procedure. What is more, the shape of the variogram can be used to determine what sort of weighting function to use.

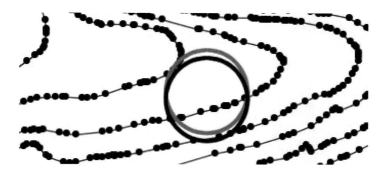

Figure 9.10 The use of inverse distance weighting with digitized contours.

Kriging is a statistical technique, and can only be used if the data being interpolated satisfies various conditions. A full explanation of the method and its limitations is beyond the scope of this book, but some references are given in the reading section for those who want to know more.

Even when kriging is appropriate, it does not overcome one of the main drawbacks of the inverse distance weighting approach which is that it does not attempt to use information about the shape of the surface. This problem becomes particularly apparent when the method is used with contour data rather than randomly scattered points. Figure 9.10 shows an example of some contours digitized from the British Ordnance Survey map of an area near Slapton Ley in Devon. The dots represent the digitized points along the contours.

The two circles indicate the points which will be selected if we use the inverse distance method with a fixed radius for selecting the points. If you look at the lighter circle, you can see that all the points selected come from a single contour, which runs on either side of the ridge crossing the area from west to east. This means that the interpolated height for this point will be the same as the height of this contour, and this will be true for all the grid points which lie within this upper contour. Even though we can see that the height along the ridge will be a little higher than this contour, all the interpolation method 'sees' is a series of points with heights attached and as a result it will produce height estimates which show the ridge as having a horizontal top, at the height of the highest contour.

The dark circle shows the points selected for a grid point which is just to the south of the first one. The circle now includes fewer points from the upper contour, but has 'captured' two points from the lower contour. This means that the interpolated height will change quite markedly from that interpolated using the dark grey circle. The effect of this can be seen in Figure 9.11 which shows the DEM produced by applying the inverse distance weighting method to a set of contours – the ridge in Figure 9.10 is in the southern part of this map. Around each contour, the interpolated heights are very similar to the contour height. Midway between each

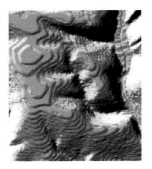

Figure 9.11 Gridded DEM produced using inverse distance weighting on contour.

contour, there is a rapid change in the interpolated heights, leading to a DEM which resembles a 'wedding cake'.

These distinctive ridges are an example of what are known as interpolation artifacts. Remember that whenever we interpolate a value there will be some error in our estimation. This error will affect any further processing we do using the data, but as long as the amount of error is relatively small the effects should not be too great. However, in the case of DEM interpolation, the errors often have distinctive spatial patterns, and form artifacts. The problem is that although the errors in the estimated height may be small in absolute terms, they may cause quite large errors in our analysis. For instance, if we calculate slope angles using the DEM in Figure 9.11, we will get values which are consistently too low around the contours, with values which are too high in between. On average the values may be about right, but if we use a map of the slope angles to identify areas suitable for ploughing or to estimate soil erosion rates, then our results will be strongly affected by the presence of the artifacts.

This illustrates that in assessing GIS algorithms, speed and efficiency are not the only concerns, and we must also be conscious of the accuracy of the results which are produced. In the case of DEM generation, we would only expect to create a DEM once, but we might use it hundreds or thousands of times, so the speed of the creation algorithm is almost irrelevant compared with the accuracy of the results it produces. We will return to point interpolation in due course, but first let us look at algorithms for creating the other major surface data model, the TIN.

9.3   ALGORITHMS FOR CREATING A TRIANGULATED IRREGULAR NETWORK

In trying to create a triangulation, we are trying to join together our sample points in such a way that they produce a reasonable representation of the

Figure 9.12 Example points to illustrate triangulation algorithms.

surface. We will look more closely at the use of the TIN model in Chapter 10, but to understand some of the issues in creating the triangulation it is useful to know that the key to the use of the TIN is the fact that the height of the three corners is known. Given this information it is possible to work out which way the triangle is facing, how steeply it is sloping and the height at any point in the triangle. In other words, the triangles do not simply represent the height of the sample points, they model the whole surface.

In the same way that we interpolated heights at unknown grid points by looking at nearby points, we should expect to produce a triangulation by joining nearby points rather than distant ones. Let us start by looking at an algorithm which attempts to produce a triangulation based on this distance measurement alone.

At first sight, this seems like quite a simple algorithm to produce. Given a set of points, simply calculate the distance between all pairs of points, then join the closest two, followed by the second closest and so on.

In the case of the points in Figure 9.12, the table of distances might be as follows:

	a	b	c	d
a		1.5	1.2	3.3
b			2.4	3.1
c				3.2
d				

Only part of the table is filled in since the distance from a to b is the same as from b to a, and the distance from any point to itself is zero. It is a simple matter to use this table to produce a triangulation – the shortest distance is between a and c so these two points are joined first, followed by a and b and so on. After five stages the triangulation will look like Figure 9.13a but at this point a problem emerges. There is still one distance in our table, but joining a and d will produce the situation as shown in Figure 9.13b in which the new triangle edge crosses one of the existing edges.

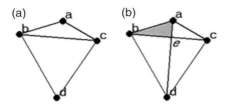

Figure 9.13 Triangulation after five (left) and six (right) stages.

We cannot allow this to happen because we do not have a height for our crossing point (labelled e in the figure) and so we would not be able to estimate any properties of the surface within the four triangles which have e as one of their corners. What happens if we do not force the two lines to intersect? We still have a set of triangles – abc, abd and acd – and we know the height at the three corners of all three. The problem is that parts of the surface are now defined by two triangles rather than one. For instance, the shaded area in the figure belongs to triangles abd and abc. If we need to estimate the height of a point within this area, which triangle would we use? Clearly then we would have to modify our distance-based algorithm so that it checks that a new edge does not cross any existing edges. Given what we already know about detecting whether lines cross, this will add significantly to the workload of the algorithm.

Obviously our simple algorithm is going to be more complicated than we thought. We cannot simply keep joining points together, we need some way of checking that we are producing a set of triangles. One possibility might be to join the closest pair of points, and then add two further lines to complete this triangle before doing anything else. We could then join the point which is closest to this triangle and complete our new triangle and carry on adding triangles to our set until we had used up all the points. This would work, but our resulting triangulation would be very dependent on where we started the process i.e. where the two closest points were. A very small change in the location of one of the points could result in a different starting point for the process and a completely different set of triangles. This suggests our method cannot be producing a very good triangulation – if it was then a small change in the position of the points should only have a small effect.

There is another weakness in this shortest distance approach. If we look at the triangulation on the left of Figure 9.13, we can see that the upper triangle is rather long and thin in comparison with the lower one. The best triangles to use for interpolation within our final TIN are 'fat' ones like the lower triangle. With thin triangles there can be a number of problems. To estimate the height of a point on the surface for instance, we find out which triangle it falls in, then fit a flat surface through the triangle corners and estimate height using this. With a long thin triangle any small errors in the location of our points, or errors caused by loss of precision will have a much greater effect on this process. We may locate the point in the wrong triangle for instance. In addition, the calculation of the triangle slope will be far more sensitive to errors for thin triangles than for fat triangles.

Just to add to the woes of this method, it may have seemed simple but it is far from efficient. To begin with we need to calculate and store all the inter-point distances. The number of distances in total is $n(n - 1)/2$. Each point is compared with every other point except itself $(n(n - 1))$ but since the distance from a to b is the same as b to a, we only need to do half the calculations. Therefore, both the computational and storage complexity is $O(n^2)$.

Creating the triangulation also has quadratic complexity. The majority of points in a triangulation lie on the junction of 3 triangles, so the number of lines is approximately $3n$ (but not exactly this, for reasons which are not important). If we have to check to see whether each line might intersect one we have already added, then for each of our $3n$ lines, we potentially have to check with up to $3n$ lines (fewer obviously in the early stages of the algorithm). However, we can see that this approach is going to have roughly quadratic efficiency in terms of both speed and storage.

All this suggests that simply joining nearby points together is not going to produce a very satisfactory result. The method which is actually used is based on the idea of producing triangles which are fat rather than thin, but it is easiest to explain this by starting with what is known at the Voronoi tessellation which is shown in Figure 9.14. Each area in this figure represents the part of the surface that is nearer to the point it contains than to any other point. In fact, this tessellation can be used for the simplest form of interpolation of all, which is sometimes used for estimating rainfall. If we assume that each point is a raingauge, then for any other point on the surface, a simple way of estimating the rainfall is to take the value from the nearest gauge. This is often done when making rough estimates of rainfall over catchments when there are very few gauges available.

What is more interesting in our case is that Figure 9.14 is another example of what mathematicians call a graph, and as in the case of the

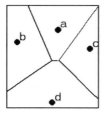

Figure 9.14 The Voronoi tessellation of the sample points.

Figure 9.15 Delaunay triangulation of the four sample points.

street/block network we looked at in Chapter 2, there is a dual to this graph which is shown in Figure 9.15.

This graph is produced by joining together points in the Voronoi tessellation which share a common boundary. When the graph is represented using straight lines between the points, then what is produced is the Delaunay triangulation, after the mathematician who first studied its properties.

One of its most interesting properties is that it can be shown mathematically that this triangulation produces the 'fattest' possible triangles. What this means is that, of all the different possible ways of triangulating a set of points, the Delaunay triangulation produces triangles in which the smallest angles within each triangle are as large as they can possibly be. Of course, long thin triangles will still be produced – when we look at the use of the Delaunay triangulation with some real data we will see this – but on average there are less of them than with other triangulations.

Another useful property is that the triangulation is mathematically defined for almost any set of points (although it is possible to define sets of points for which the triangulation is not uniquely defined). It does not matter how we calculate it, it will always be the same (provided we do the calculations correctly!). What is more, if the points change slightly, the triangulation only changes slightly too, and only in the area near the altered point. Finally, it can also be proved mathematically that none of the Delaunay triangles edges cross each other, so there is no need to include a check for this in our algorithm.

Defining the triangulation is one thing – but how do we produce it in the computer? The answer lies in the use of some of its mathematical properties. Consider the two triangulations of the four points in Figure 9.16. The one on the right is clearly preferable to the one on the left, since both triangles have minimum angles which are larger than the minimum angle of the 'thin' upper triangle on the right.

What is interesting is that this can be tested mathematically. Assume we have just added the edge bc to the triangulation and want to test if it is a proper Delaunay edge (remember that the Delaunay triangulation is mathematically defined for a set of points – our task is to find it). What we do is take one of the triangles that bc belongs to and draw the circle which goes through all three points. Another of the properties of the

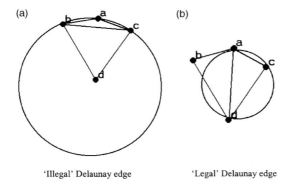

'Illegal' Delaunay edge   'Legal' Delaunay edge

Figure 9.16 Edge flipping to produce a Delauney triangulation. (a) 'Illegal' Delaunay edge (b) 'Legal' Delaunay edge.

Delaunay triangulation is that a circle drawn through three points which are joined in the triangulation will not contain any other point. In the example on the left this is not true, so the edge we are testing cannot belong to the Delaunay triangulation for this set of points.

The edge in question belongs to two triangles, and together these form a four sided figure. If we 'flip' the edge to make it join the other two corners, then we will produce a legal Delaunay edge as shown in Figure 9.16b. Note that once we flip the edge, the definition of the triangles changes, and hence so does the circle, which now joins the three points acd, but excludes point b. This test and edge flipping will always work as long as all the other lines apart from the one we are testing belong to the Delaunay triangulation. This is what gives us the key to an algorithm for computing the full triangulation. Rather like the algorithm we used for sorting, the approach we take is to begin by triangulating just three points, and then adding extra points one at a time. Each time we add a point, we find out which of the current triangles it falls in and draw edges from the new point to the three corners, which will produce three new triangles as shown on the left of Figure 9.17.

The new edges must all be legal Delaunay edges (for reasons explained in a moment). However, adding these edges may affect edges in the neighbouring triangles, as shown on the right of Figure 9.17. The edge of the original triangle has become illegal with the addition of the new edges, and needs to be flipped to the position shown by the dotted line. This in turn will affect the two edges shown in grey, which will have to be checked in their turn.

But how do we pick the three points to start this process off? In fact it does not really matter. When we only have three points, the triangle that joins them is by definition the Delaunay triangulation of those three points. As we add each successive point, the edge flipping will ensure that we always have a Delaunay triangulation of the current set. However, as we

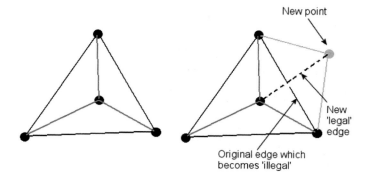

Figure 9.17 Addition of a point to a Delaunay triangulation.

proceed some of the points which we add will be inside existing triangles, and some will be outside the triangulated area. To avoid having to distinguish between these two cases, we can start by producing a large triangle which completely encloses all the points. This means that every point we add has to be inside one of the existing triangles, simplifying the implementation of the edge testing procedure.

Because there is no need to calculate the huge array of distances which we needed for the first algorithm, the storage requirements are much more modest, and in fact turn out to be O(n). The processing efficiency is O($n \log n$) which is the same as the sorting algorithm and for similar reasons – we need to process n points and for each we need to find out which of the current triangles it falls into. Now that we have discovered how to calculate the Delaunay triangulation, let us consider some of the issues in using it to create a model of the terrain surface.

9.4   GRID CREATION REVISITED

We saw in Section 9.2 that there are two desirable properties for algorithms which perform point interpolation – they should use information from nearby points, and they should attempt to model the surface. The methods described in Section 9.2 all satisfied the first condition, but did not really try to satisfy the second. In this section, we will look at some other algorithms which do try to tackle the second issue.

The first is the TIN model. At the heart of the construction of a TIN is the idea that the points which are joined are natural neighbours of one another. Hence, the Delaunay triangulation provides a different answer to the question of which points should be considered to be 'nearby'. It also provides a means of modelling the surface, since the triangulation produces

Figure 9.18 How the TIN models the surface.

a set of triangular patches which completely cover the area. Since we know the height at the corners of the triangles, we know which way they are facing, and so what we have is a model of the surface, as shown in Figure 9.18. This shows part of the valley which we saw in the previous chapter, modelled using a TIN rather than a gridded DEM. The contours have been used as the data source again, and the triangles which model the surface are clearly visible in this shaded view.

In order to estimate height at an unknown point all we have to do is find out which triangle it falls in and then estimate the height of the triangular patch at that point. This is relatively simple, but has a major drawback, in that the model is one in which flat triangles meet at sharp edges. This means that our estimates of height will change abruptly along these edges, which is unrealistic. One solution to this is to assume that the triangular faces are not flat but curved in such a way that they meet smoothly along their edges, with no abrupt changes of slope. This can be done by fitting mathematical surfaces called polynomials to each triangular facet. These are most easily explained by taking a simple two dimensional case.

Figure 9.19 shows three simple curves:

$$z = 60x$$
$$z = 10x^2$$
$$z = x^2 - x^3$$

Each curve is a polynomial formed by combining successively larger powers of x. The linear function is a straight line, the quadratic function is a U-shaped curve and the cubic function is like an S on its back. As the graph shows, each additional power of x effectively adds another 'bend' to the curve. The same principle applies to 2D polynomials, where the height of the curve (z) is a function of powers of the two horizontal dimensions, x and y, and the resulting curve forms a 2D surface rather than a one-dimensional

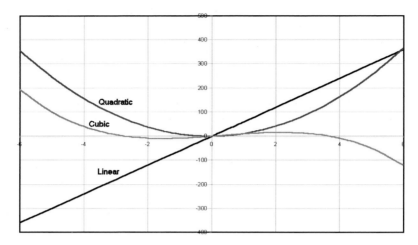

Figure 9.19 The graphs of polynomials of different order.

line. The larger the powers of x and y, the more curved the surface will be, and by using quintic polynomials (which include terms up to power 5) it is possible to fit a surface to each triangle in a TIN such that it meets the surfaces in the neighbouring triangles smoothly.

The difference is best illustrated with a real example, which shows the interpolation of heights from the contours in Figure 9.20. In the left hand figure, a TIN has been fitted to the contours, and a gridded DEM produced assuming each triangle is a plane surface. In the right hand figure the interpolation to the DEM has been performed by fitting a quintic polynomial to each triangle. The results are slightly smoother on the slopes, but

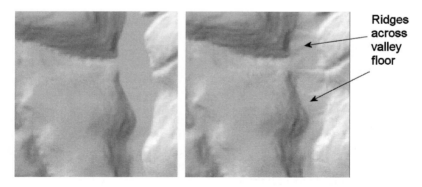

Figure 9.20 Gridded DEM produced from a TIN by linear (left) and quintic (right) interpolation.

the linear method produces a better model of the flat valley floors, whereas the quintic method produces a series of small ridges crossing the valley.

This would therefore seem to provide an ideal way of interpolating a gridded DEM since it meets both of our criteria. However, there is one weakness with this approach. The idea behind using a triangulation for representing the terrain surface was that the triangles could be used to actually model features of the surface. For instance, triangles edges could run along ridges and down valleys, large triangles could be used where the terrain was smooth, and smaller triangles where it was rough. It is possible to construct what is known as a constrained Delaunay triangulation in which certain edges are forced to be part of the triangulation, even if they break the mathematical rules defining a true Delaunay triangulation.

This idea is fine in theory, but in practice it is dependent on having samples of elevation along the key features of the landscape which can then be used to build the triangulation and this is not often the case. It is certainly not the case with contours and in fact there are several well known problems with fitting a standard Delaunay triangulation to contours. Figure 9.21 shows a map of the aspect of each of the triangles in Figure 9.20. The triangles which are white are all flat, because they join points which are on the same contour. As the figure shows, this happens along the ridges, and down the valleys, producing what appear to be a set of terraces in the landscape.

The problem arises because the standard Delaunay triangulation is based solely on the x and y values of the points and pays no attention to the z values. As already described in this section, one way to avoid these problems is to provide additional data points along valleys and ridges, which help define the surface properly at these points.

This reinforces the point made in Section 9.2 that in the case of algorithms for creating representations of surfaces the key issues tend to revolve around the quality of the final surface, rather than the efficiency of the algorithm. One of the problems, which is the subject of a good deal of current research,

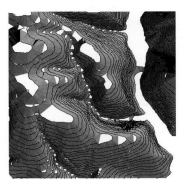

Figure 9.21 Aspect of triangles in Figure 9.20 – areas shown white are flat.

is that there is no simple, objective measure of this quality equivalent to the big-O notation for assessing the quality of different algorithms.

FURTHER READING

Units 38 and 39 of the NCGIA Core Curriculum describe the gridded and TIN models of surfaces respectively (http://www.geog.ubc.ca/courses/klink/gis.notes/ncgia/toc.html). The vector/raster debate, which was so important in the early days of GIS development, has a counterpart in the debate about the relative merits of the gridded DEM and the TIN for representing surfaces. The TIN was developed partly as an answer to the perceived weaknesses of the gridded model, as explained in some of the early work by Peucker and colleagues (Peucker *et al.*, 1978; Mark 1975, 1979). More recently, Kumler (1994) has undertaken a comprehensive comparative review, concluding that although the TIN seems intuitively a better representation of the landscape, many algorithms produce better results on a gridded DEM.

One of the few systems to use a contour-based DEM is TAPES-C from the group at the ANU in Canberra – more details on their home page (http://cres.anu.edu.au/outputs/tapes.html) and in the book edited by two of the group (Wilson and Gallant 2000).

The description of the Delaunay triangulation algorithm in this chapter is partly based on the material in Chapter 9 of De Berg *et al.* (1997). Chapter 7 of the same book describes the Voronoi tessellation. The Voronoi tessellation and Delaunay triangulation both have uses outside surface interpolation in GIS. For instance, Geographers know the polygons which form the Voronoi tesselation as Thiessen polygons, after a meteorologist who 'invented' them as a means of estimating areal rainfall amounts based on values from raingauges. A good starting point to find further information on the range of applications is Chris Gold's Voronoi Web Site (http://www.voronoi.com/). Both Gold (1992) and Sibson (1981) describe an interpolation algorithm based on the Voronoi tesselation.

Interpolation has been the subject of a good deal of discussion. Unit 41 of the NCGIA Core Curriculum (http://www.geog.ubc.ca/courses/klink/gis.notes/ncgia/toc.html) gives a good overview of a range of methods. Burrough and McDonnell (1998) and Lam (1983) also reviews a wide range of different techniques – Burrough and McDonnell (1998) give a good description of kriging in particular. A more recent review is available online at review is available online at (http://www.agt.bme.hu/public_e/funcint/funcint.html).

10 Algorithms for surfaces

Now that we have seen how we can create the two main surface models, it is time to consider their use in answering queries about real world surfaces. Again it is important to remember that all our models are just that – models of the true surface. The degree to which the model is an accurate representation of the real surface will affect the degree to which the answers we get when using it reflect characteristics of the real surface.

We have already seen that many of the methods for interpolating a gridded DEM from elevation data give rise to artifacts which will clearly affect our results. In this chapter, we will see whether the results are also affected by the algorithms we use.

10.1 ELEVATION, SLOPE AND ASPECT

The most fundamental property of a surface is its height. If the point we are interested in happens to be one of the grid points in a gridded DEM, or one of the triangle corners in a TIN (i.e. one of the original sample locations) we can answer the query straight away. In general, of course this is unlikely to be the case, and we are going to need a method for estimating height at any point on the surface. We have already covered the point interpolation problem, because this is how we created the gridded DEM from the original sample data. However, in that case we did not have a surface model – simply a collection of points – and we can use the fact that we have a model to simplify the point interpolation process. In particular, both the DEM and the TIN tell us what the height is at nearby points on the surface, so that all we have to do is decide how best to use this information to estimate the height at our unknown point.

In the case of the TIN, this is a two stage process. First, we need to identify the triangle which our point falls into. Second, we use the information about this triangle, and possibly its neighbours, to estimate height at the unknown point. Given a set of triangles and a point, we could use one of the point in polygon algorithms which were described in Chapter 4 to answer the first part of the query. However, these were designed to work with polygons of any shape and size. In this case all our polygons are triangles, and all the polygon's sides are made up of single straight lines and we can use this information to speed up our point in polygon algorithm.

Triangles are what are called convex polygons. Informally this means that they do not have any indentations. Formally, it means that any two points within the polygon can be joined by a straight line which does not cross the boundary. Other examples of convex polygons are the regular geometrical shapes such as the square, rectangle, hexagon etc. This simplifies the point in polygon algorithm because a ray from a point will only cross the boundary once (if the point is inside) or twice (if the point is outside). In fact rather than use the normal point in polygon test, we can exploit this property to produce a special purpose test.

If a point falls inside a triangle then its Y coordinate must fall in between the Y coordinates of two of the triangle corners and likewise its X coordinate. This means we can use a very rapid test to exclude from consideration all triangles which lie entirely above, below or to one side of our point. In many cases, this will leave a very small number of triangles, which can be tested using the normal ray method. This approach can be speeded up even more if we use a spatial index of the sort described in Chapter 8 to store the points in the triangulation, since we can use the index to identify nearby points and hence potential enclosing triangles.

Once we have found the correct triangle, we can find how steeply it is sloping and which way it is facing from the coordinate of the three corner points. The mathematics of how this is done is beyond the scope of this book, but an explanation of why this is always possible is useful, and has a bearing on some of the material which will be covered further on. The problem is to fit a 2D, plane surface through three points in space. Let us start with the simpler problem of fitting a one-dimensional line through two points in space. To simplify things even further, we will make one of the points the origin of our coordinate system (the point with coordinates $0, 0$) as shown in Figure 10.1.

The general equation for a straight line, which we used in Chapter 3, has the following form:

$$z = a + b \cdot x \qquad\qquad 10.1$$

This has two unknown parameters – a and b. The value of b tells us how steeply the line slopes – for each increase in the value of X by 1, what will be the increase in the value of Z. The value of a is what is called the intercept – when X has a value of 0, what value does Z have? In this case, we know that when X is 0, Z is also 0, so in fact we only have one unknown quantity – b – and our equation will be

$$z = b \cdot x \qquad\qquad 10.2$$

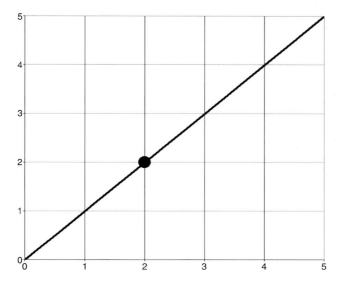

Figure 10.1 Graph of $z = ax$.

We can see from looking at the graph, that the value of b in this case is 1. When x increases by 1 unit, z also increases by 1 unit. But if we had to work this out given the location of some points along the line, how many points would we need? The answer is just one. To work out the slope we need to work out how far the line rises in Z for a given rise in X and we can work this out from the position of one point on the line.

Now consider the line in Figure 10.2. This no longer goes through the origin, and so the equation of this line is:

$$z = b \cdot x + a \qquad\qquad 10.3$$

The line still has a slope of 1, but it is higher up the y axis – 1 unit higher up in fact, so b has a value of 1. This line has one more unknown parameter compared with the first one – a – and so we will need a second point along the line so that we can calculate this second parameter. Now imagine that the line is extended to a third dimension, so that it becomes a flat surface in 3D space. If we only have our two known points, we are not going to be able to tell exactly where this surface goes. Imagine taking a sheet of card and laying it along the straight line. We could spin the card through 360°, with it still lying along the line. However, as soon as we have a third fixed point, this will fix the card into position in space. The equation of the plane will be

$$z = a \cdot x + b \cdot y + c \qquad\qquad 10.4$$

We now have three unknowns – a, b and c – and we have seen that we can work these out as long as we have three points on the plane. Since we have

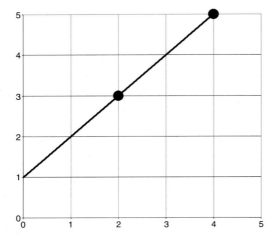

Figure 10.2 Graph of $z = ax + b$.

the height at the three corners of our TIN triangles, we can always fit a plane surface through them.

Once we have the equation of the plane passing through the triangle of the TIN, we can use the parameters to tell us the characteristics of the surface. Elevation is obtained simply by putting the x and y coordinates of our point into the equation. Slope and aspect can be calculated from the values of a and b which tell us how steeply the plane is sloping in the X and Y directions respectively. Note that a and b do not give us a slope in degrees, but as a proportion. If we look again at the line in Figure 10.2, the slope of this line can be calculated as follows:

$$\text{slope} = \frac{z_2 - z_1}{x_2 - x_1} \qquad\qquad 10.5$$

assuming the two points have coordinates (x_1, z_1) and (x_2, z_1). This calculation produces a value of 1 in this case, which is the tangent of the slope in degrees. The angle is therefore the arctangent of 1 which is $45°$. Slopes calculated as the ratio between the fall in height divided by the horizontal distance in this way are sometimes multiplied by 100 and reported as a percentage. This is often seen on road signs warning motorists of steep gradients on roads.

Given values for a and b, the following formulae will calculate the maximum slope of the plane (also expressed as a proportion) and the direction it faces (as an angle with north as zero)

$$\text{slope} = \sqrt{a^2 + b^2}$$
$$\text{aspect} = \tan^{-1}\left(\frac{-a}{b}\right) \qquad\qquad 10.6$$

This assumes that the TIN surfaces are modelled as planes. However, it may be more realistic to assume that the triangle surfaces are curved as we saw in Chapter 8. Many GIS packages use quintic polynomials (which contain terms up to powers of 5) for this purpose. Since these contain higher order terms, we know that they can model surfaces which curve in several directions. In fact, using quintic polynomials, it is possible to ensure that the curves defining neighbouring triangle surfaces meet smoothly along the triangle edges, which removes the angular appearance of DEMs produced using plane triangles.

However, there will be more than three unknowns for the polynomial equations, since as well as parameters defining the slope in x and y, we also have parameters defining the amount of curvature in various directions. We cannot fit these surfaces using just three points, so we have to use points from the neighbouring triangles in addition. This makes intuitive sense. We are now modelling our triangle, not as a flat surface, but as a surface with a bulge in it. The only way to find out whether the triangle surface bulges a little or a lot is to look at the slope and direction of the neighbouring triangles – if these are very similar to the one we are interested in, then the surface must be fairly planar at this point, but if they are very different it must be quite curved.

Finding the elevation, slope and aspect at any point on a gridded DEM is the same in principle as with a TIN but the details of the calculations are very different. The first stage is relatively trivial. With a TIN we need a special algorithm to find out which triangle our query point lies in. With a grid, the points are spaced in a regular layout, so it is very easy to find out which grid cell the query point falls in – the coordinates of the point tell us directly. However, the cell itself will not help us answer our query as the triangle did. The cell is simply the square area immediately surrounding the grid point – it does not have a slope or direction as the triangle did. In order to estimate out surface parameters, we need to use the information from the neighbouring points. There are actually several ways in which this can be done.

The two equations for slope and aspect above are based on the slope in the x and y directions. In a grid, these are potentially very simple to estimate – for instance in the grid in Figure 10.3, the slope in the y direction can be calculated from using the elevations of the points in the rows above and below the central point.

$$Y \text{ slope} = \frac{B - H}{2d}$$

$$d = \text{grid spacing.}$$

10.7

A similar calculation using the points from the neighbouring columns will give the slope in the x direction, and the two values can simply be inserted in the equations. By doing this we are effectively assuming that the slope and

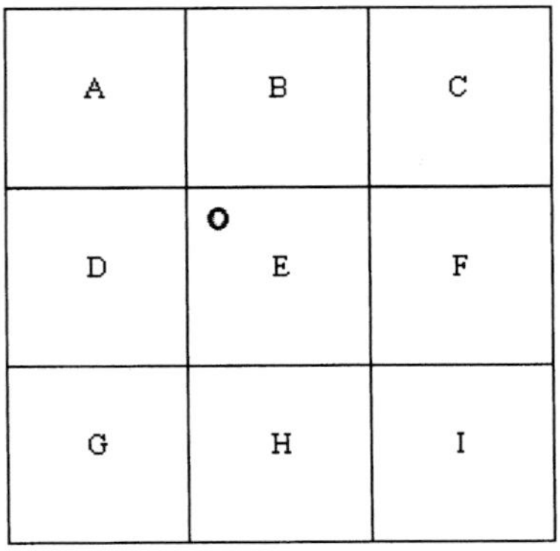

Figure 10.3 Location of cells in 3 × 3 window used for estimating surface
properties at a point.

aspect at the unknown point are the same as slope and aspect at the nearest
grid point (i.e. the central point in Figure 10.3). Applying the same logic to
elevation, we can assume that height at any point is the same as the height of
the grid cell it falls into. There are several problems with this approach. One
is that the results are very dependent on the accuracy of the heights in each
of the grid points – a slight error in the height in pixel B for example will
produce an error in the estimate for b. The same was true for the TIN
method, since the elevation, slope and aspect were all essentially calculated
from just three data points – the heights at the corners of the enclosing
triangle. The difference was that these heights were measured, whereas the
heights in a gridded DEM are all interpolated estimates which means that in
addition to any measurement error they will contain errors introduced by
the interpolation process. For this reason, the equation used to estimate
slope in the X direction at the central grid point is usually:

$$Y \text{ slope} = \frac{(A + 2B + C) - (G + 2H + I)}{8d} \qquad 10.8$$

Thus rather than simply looking at the difference between B and H, the
difference between the average height of the points in the rows above and
below is used.

Another method altogether is to fit a surface through the points in the
neighbourhood, as was done with the TIN. This has the advantage that
elevation, slope and aspect can be calculated at any point within the
neighbourhood, not simply the central point. With the nine data points in
a local window, such as in Figure 10.3, it is possible to fit a surface with 9
parameters, and some algorithms do this. However, it may well be better to

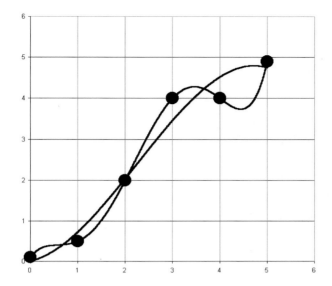

Figure 10.4 Line through three points.

fit a somewhat simpler surface. To understand why, let us return to the example of fitting a straight line to a series of points.

Figure 10.4 shows a series of 6 data points which are estimates of elevation along an imaginary hillslope. It is possible to fit a curve which will go through all six points exactly as shown – this is a fifth order polynomial. The graph shows that in order to pass exactly through each point the curve has to bend quite sharply in places. If we used this curve to estimate height or slope at any point along the line, then some of our estimates will seem rather odd – for instance towards the right hand end of the graph, the hillslope will apparently have a negative slope, and appear to be facing in completely the wrong direction.

However, we can also fit a line which does not pass through every point exactly, but tries to capture the overall trend of the data points. In this case, a third order polynomial has been used. We can see that at the upper end of the slope the line traces a more plausible shape between the actual data points. This line will also be much less sensitive to small changes in the data points. The same principle applies to fitting a surface through the points in our 3 × 3 window in Figure 10.3. A surface which goes through all nine data points will be far more sensitive to errors in the DEM, than a simpler curve.

As an illustration of the difference which these factors can make in the final results of using a DEM, consider the two diagrams in Figure 10.5. which are both maps of aspect derived for the same area, using the same contours as the original data. The aspect values have been classified to

(a)   (b)

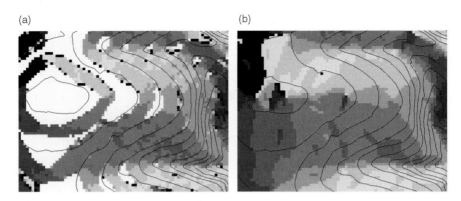

Figure 10.5 Slope aspect estimated using two different DEMs.

represent the main compass directions (N, NE etc.) which are represented by the different grey shades. Both DEMs have captured the main changes in aspect reasonably well, distinguishing between the three sides of the central ridge for example.

The DEM as shown in Figure 10.5a has been produced using an Inverse Distance Weighting algorithm, and this produces flat areas (shown in white) on the hilltop, and in between some of the contours. In contrast, the DEM as shown in Figure 10.5b has been produced using a TIN, with quintic interpolation, and this produces reasonable estimates of aspect for many of these areas (although the values on the hilltop seem strange in places). The DEM in Figure 10.5a also shows the effect of using the poorer equation for estimating aspect (Equation 10.5). Since this is sensitive to differences in the height of individual grid points, it produces individual aspect values which are markedly different from those in neighbouring points – these are most clearly seen in the isolated black pixels which occur over the map, especially along the fringes of the flat areas. Figure 10.5b has been produced using Equation 10.6, and this, combined with a smoother initial DEM, produces a map in which aspect varies more gradually across the terrain.

10.2   HYDROLOGICAL ANALYSIS USING A TIN

One of the major uses of surface models in GIS is the analysis of the flow of water across the landscape. Important tasks such as estimating the volume of water flowing in rivers and accumulating in reservoirs, assessing the risk of flooding and estimating soil erosion rates all require a knowledge of how water will flow across the terrain.

Central to this analysis is modelling what direction the water will take across the surface. Once this can be done for any point on the surface, all the other things we need to know can be based on this. For instance, when flow paths come together it can be assumed that these will become river courses. Given any point on the landscape we can trace the flow directions upslope to define the area which drains to that point – this area is called the catchment in the British hydrological literature, the watershed in the American literature.

Since we know that water is driven by gravity, it will always follow the steepest path downslope, and so we can use this to model flow across our digital landscapes. Figure 10.6 shows part of a TIN representing a very simple river valley. It should be clear from an inspection of the heights of the triangle vertices that the triangles converge towards a series of edges which define the river valley which flows from right to left across the figure. This suggests one method of directing the flow across the TIN – assume that water will flow across each triangle to the lowest edge, and then flow along the edges, which effectively define the stream channels. This will work for triangles which directly border major streams, such as D and F in Figure 10.6. However, consider what this would mean for triangle C – water would flow to the lower edge, and then would all be forced to flow along this edge to the vertex marked 40, where it would meet the main valley.

This simple approach would lead to some very strange looking directions for drainage channels, and it is clear that to avoid this we will have to allow water to flow from one triangle to another across edges. This means for instance that water from C will flow across F and hence to the valley.

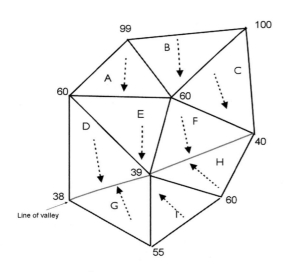

Figure 10.6 Flow across an imaginary TIN representing part of a river valley. The TIN edges representing the bottom of the valley are drawn in grey.

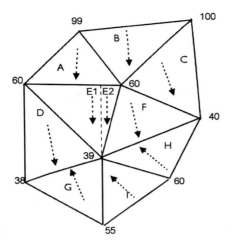

Figure 10.7 Subdividing flow across triangle E.

However, even this will produce some strange results. For instance, triangle E meets the stream at a vertex rather than an edge, and so flow from this triangle is forced to go into one of its neighbours (D or F) in order to reach the valley. This means that water falling in the upper right hand corner of E is predicted to flow across both E and D, whereas water which falls nearby on the top of F will take a more direct route to the stream. One way to avoid this would be to subdivide the flow from triangle E into two parts as shown in Figure 10.7, with some being passed to D and some to F. This means that we will have to modify the original triangulation to include this new subdivision of triangle E.

These problems arise because of our assumption that channels can only exist along traingle edges. However, if we allow channels to direct water across the triangles, we run into further difficulties. In Figure 10.8, the flow of water across F to the valley at the bottom is shown. If we view this flow as a single channel running down the triangle, then we have two channels meeting at the point f. In a real river network channels are usually separated by a ridge, so two channels meeting on a plane surface like this is not very realistic. The implication is that there should be a ridge between them, as shown by the dotted line in Figure 10.8b. However, this produces a new triangle (labelled F2). When we model flow across the surface of F2, we have the same problem as with F, and we will need to add a further ridge, producing another small triangle, and so on *ad infinitum*.

One solution to these problems is to use a triangulation which has been carefully constructed so that all the ridges and valleys in the landscape are properly modelled by triangle edges. In practice many triangulations are produced by applying a standard Delaunay triangulation routine to a set

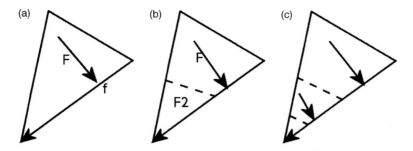

Figure 10.8 The problem of infinite regress in defining drainage channels on a TIN.

of existing elevation data, and there can be no guarantee than the resulting triangulation will be a good model of the surface. Indeed, if the original data were from contours, there is good reason to believe that the triangulation will misrepresent key features of the landscape such as ridges and valleys.

10.3 DETERMINING FLOW DIRECTION USING A GRIDDED DEM

The majority of GIS-based hydrological modelling uses the gridded DEM rather than the TIN. The main reason for this is the simplicity of handling the simple array data structure of a gridded DEM, compared with the complex TIN data structure. Some of the features of using the gridded DEM will be described in this section – for simplicity the term DEM will be taken to mean a gridded DEM throughout.

The key to modelling the flow of water across the surface is to determine the direction in which water will flow from each DEM cell, which is done by considering the local 3×3 neighbourhood surrounding each grid point. The simplest method is to assume that water will flow to the neighbouring cell which has the steepest slope from the central cell. This will generally be the neighbour with the lowest elevation, but allowance has to be made for the fact that the diagonal neighbours are slightly more distant from the central cell than the others.

At first sight this seems a reasonable approach, since it represents the idea that water flows in the steepest downslope direction. However, it is not difficult to think of situations where the downslope grid point does not represent the main downslope direction from the centre point. Figure 10.9 shows a portion of a DEM which has an aspect of approximately 177° (i.e. very nearly due south). However, flow from the central cell will be to the south-east neighbour, a direction of 135°.

100	99	98
80	79	78
60	59	48

Figure 10.9 Imaginary DEM. The terrain is sloping almost due south, but flow from each cell will go to the south-eastern neighbour.

Strictly speaking of course, water will not flow from point to point across the surface, and in considering a local neighbourhood of grid points, such as in Figure 10.9, it is the flow of water from the cell surrounding each point which is being modelled. More sophisticated approaches are therefore based on two ideas:

1   That the flow from the cell all travels in the same direction, but that this is not constrained to be directly to one of the neighbouring cells. This means that unless flow is directly to a neighbouring cell, water may be distributed among several neighbouring cells.
2   That since the flow originates from an area rather than a point, it may not travel in the same direction. In the case of flow from a hilltop, this might result in water being distributed to all of the neighbouring cells for instance.

The different flow direction algorithms will differ in the complexity of the calculations which need to be done for each grid point, but in terms of overall complexity they will all be $O(n^2)$, since the calculations are done once for each grid point. A more telling comparison is that the single flow direction method is very sensitive to small errors in the DEM, since a slight error in a single cell can completely alter the predicted direction of flow.

Once the direction from each cell has been estimated, the next step is to estimate for each cell, how many other cells are 'upstream' of that point on the landscape. Cells which have a large catchment area can be assumed to be on the stream network, so a simple thresholding operation on the accumulated flow count will produce a predicted stream network.

The direction of flow from each cell is stored as a code number in that cell. Since there are eight possible directions, it might seem sensible to store them as numbers from one to eight. In fact, in many systems, the numbers used are as shown in Figure 10.10.

32	64	128
16		1
8	4	2

Figure 10.10 Codes used for directions out of a DEM cell.

The reason for this apparently strange set of numbers becomes clearer when they are each written out in binary form as shown in Figure 10.11. This shows that each number has a single 1 bit, with the rest of the bits being set to zero. There are two advantages to this. First, it is possible to store more than one flow direction from each cell. For instance, if water flows to both the south and the south-west, the code would be $4 + 8 = 12$ or 00001100 in binary. Because each of the eight codes only uses one bit, however many of them are added together, the new number will be unique.

The second advantage of this numbering scheme is that it allows a very efficient method of using the numbers to determine flow from cell to cell. The key to using the flow directions is to be able to tell, for any given cell, which of its eight neighbours contribute flow into it. For example, if the central cell in Figure 10.12 receives flow from the cell in the north-west corner, then the flow direction code in this north-west neighbour will be 2 – flow to the south-east. To determine how many of the neighbouring cells are sources of water it is therefore necessary to look at each in turn and test whether it has the value shown in Figure 10.12.

The simplest way to do this is with a direct test for the number. In the pseudo-code which follows, flowdir is an array which stores the flow directions in the eight neighbours of the cell being considered. Flowdir(nw) is the value in the northwest neighbour, flowdir(n) the value in the neighbour to the north and so on. To count how many neighbours contribute flow to the current cell, the following set of tests would be used

```
if (flowdir(nw)=2) n=n+1
if (flowdir(n)=4) n=n+1
if (flowdir(ne)=8) n=n+1
```

etc.

Flow direction code – decimal	Flow direction code – 8 bit binary
1	00000001
2	00000010
4	00000100
8	00001000
16	00010000
32	00100000
64	01000000
128	10000000

Figure 10.11 Flow directions codes in decimal and binary.

However, this method will become very cumbersome if we allow multiple flow directions from a cell. In this case, flow from the north-west cell could be represented not only by a value of 2 (flow to the south-east) but also by a value of 6 (flow to both south and south-east) for example. In fact, of the 256 possible flow direction values which could be stored in the north-west neighbour, half of them would include flow to the south-east. Since the same is true of all eight neighbours, using the simple approach shown above would require 1024 separate if tests.

2	4	8
1		16
128	64	32

Figure 10.12 Codes which represent flow into a cell from each neighbour.

Bit	7	6	5	4	3	2	1	0
Value	0	0	0	0	0	0	1	0

Figure 10.13 Storage of 2 in 8 bit byte.

A more rapid method of performing this series of tests uses the fact that each direction is represented by a different bit in the flow direction code as shown in Figure 10.11. No matter what combination of directions are represented in the flow direction code, if there is flow to the south-east the appropriate bit will be set to 1. The lower row of Figure 10.13 shows the number 2, which represents flow solely to the south-east, stored as an 8 bit binary integer. To help with the explanation, the individual bits have been numbered in the upper row of Figure 10.13. If there is flow to the south-east, then the number 1 bit will be set to 1 – if not it will be a zero. If we can make this the leftmost bit in the byte, then there is a very rapid test we can use to see if it is a one or zero, because in a signed byte, the leftmost bit is used to indicate whether a number is positive (leftmost bit 0) or negative (leftmost bit 1).

We have already seen in Section 7.3, that shifting bits within a byte is a very quick operation, so our test for whether a particular code includes flow to the south-east is simply:

```
if (leftshift(flowdir(nw),6) is negative) n=n+1
```

Leftshift is a function which shifts all the bits in a byte left by the specified amount – 6 in this case. The same logic is applied to the other neighbours:

```
if (leftshift(flowdir(n),5) is negative) n=n+1
if (leftshift(flowdir(ne),4) is negative) n=n+1
```

etc.

This does not take any more time than the original code which tested for single flow directions – indeed it may be a little quicker because left shifting and testing whether a number is positive or negative are both rapid operations.

10.4 USING THE FLOW DIRECTIONS FOR HYDROLOGICAL ANALYSIS

The flow direction code is used in two of the most important surface operations – determining the size of the area upstream of a point, and labelling the

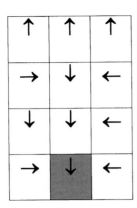

Figure 10.14 Flow directions in an imaginary DEM.

cells which lie in this watershed. Imagine that we wish to identify all the pixels which are upstream of the shaded pixel shown in Figure 10.14.

The method suggested by Marks *et al.* (1984) is an elegant one which makes use of the idea of recursion which was used for traversing the quadtree data structure in Section 7.3. In outline, the algorithm is very simple:

```
1  procedure wshed(pixel-id)
2  for each of 8 neighbours
3    if IN watershed ignore
4      if neighbour is upslope then
5        mark pixel as IN
6        call wshed(neighbour-id)
```

The algorithm starts with the watershed outflow point. In the pseudo-code, it is assumed that there is some way of labelling each pixel so that it is possible to keep track of which pixels have already been found to be in the watershed. In practice, this would probably be done by using an array of the same size as the DEM itself, so that the pixel-id referred to would be the row and column number of the pixel in question. The algorithm visits each of the eight neighbours in turn. The first check is whether the pixel has already been dealt with, since pixels which are inside the watershed will on average be visited eight times. The next step is to check whether this neighbour flows into the original pixel which is where the checking of the flow direction code comes in. If the neighbour does flow to the original pixel, it is marked as IN. At this point, the algorithm proceeds to check the eight neighbours of the pixel it is currently in by calling itself.

If the algorithm starts with the neighbour to the north, and proceeds clockwise, then after three iterations the result will be as shown in Figure 10.15.

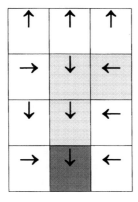

Figure 10.15 Watershed after 3 iterations of algorithm.

An outline of the sequence of operations might look like this.

```
1   Call wshed from start point
2   N neighbour IN watershed
3     Call wshed from N neighbour
4     N Neighbour IN watershed
5       Call wshed from N Neighbour
6       N neighbour not IN
7       NE neighbour not IN
8       E neighbour IN
9         Call wshed from E neighbour
```

This algorithm is very efficient in processing terms, because it only considers pixels which are in the watershed, or immediately adjacent. Each will be considered a constant number of times (eight) so the processing complexity is O(k) where k is the number of pixels in the watershed. Note that the efficiency of this algorithm depends on the size of the output rather than the input, which is why the letter k is used instead of the letter n. If the watershed being identified has 50 pixels, it will take the same number of operations to identify these 50 pixels in a DEM of 100, 1000 or 100 000 pixels. The algorithm is also very elegant and simple. The only drawback is that it might have large memory requirements. Every time wshed makes a call to itself, the computer has to set aside extra memory to store the information which identifies which pixel is the 'current' one. For example, when called from the starting pixel, this might be identified as row 0, column 1. When the north neighbour is identified as belonging to the watershed, wshed is called again, this time with a new row and column number (1,1). These will have to be stored separately from the original 0,1 row and column numbers, because at some stage the program has to come back and finish off going round the neighbours of this original point. This means that every time wshed calls itself recursively, two extra memory locations are needed. The maximum amount of extra memory required will depend on the size of the watershed, but in the worst case could be O(n), if the maximum dimension of the watershed is of the same order as the size of the raster layer.

Once the watershed has been determined, it is a very simple matter to determine its size, by counting the number of pixels which are IN, and this could be added to the above algorithm very easily. However, it is often necessary to determine what is called the flow accumulation, which is effectively the size of the watershed upstream from every pixel. If this were done by running the watershed algorithm for each pixel in the layer, this would have O($k \cdot n^2$) complexity (remembering that we are taking n as the number of pixels along one edge of a raster layer, so that each layer has n^2 pixels).

The problem is that we cannot calculate the flow accumulation simply by looking at the immediate neighbours of each pixel since each of these may receive flow from an unknown number of other pixels. However, applying the watershed algorithm to each pixel is wasteful, because neighbouring pixels will have watersheds which are very similar so in effect we will be processing the same set of flow directions several times. An elegant solution to this problem suggested by Mark (1984), is to first sort the pixels into descending order of height. A pixel can only receive flow from pixels which are higher than it. If we process this new array in height order, it should be possible to deal with each pixel only once.

To see how this might operate in practice, we assume that we have already have a raster grid containing the pixel heights and a second grid containing the flow directions, as shown in Figure 10.16. The directions have been shown as arrows in the diagram, for clarity, but in practice the numerical codes described earlier would be used of course. The output of the algorithm will be a third grid in which each pixel contains the number of pixels which flow into it, which is also shown in Figure 10.16.

Let us assume for a moment that we have taken the heights from the DEM and stored them in a second array, which we have sorted into descending height order. The algorithm for assigning flow would then be as follows:

```
1   Array HEIGHTS[1..N]
2   Array FLOWS[1..N]
3   Set FLOWS to 0
4   for i = 1 to N
5     Add FLOWS[i]+1 to FLOWS[downslope neighbour]
```

Another array, called FLOWS, is set up to contain the flow acumulation value for each pixel. As each pixel is processed, it contributes 1 unit of flow

147	127	124	137	167
131	110	108	122	153
114	94	92	107	140
98	79	77	93	126
82	63	62	79	114

↘	↓	↓	↙	↙
↘	↓	↓	↙	↙
↘	↓	↓	↙	↙
↘	↓	↓	↙	↙
↘	↓	↓	↙	←

0	0	0	0	0
0	2	2	1	0
0	4	5	1	0
0	6	8	1	0
0	8	11	1	0

Figure 10.16 Upper: heights for a small part of a gridded DEM. Middle: flow directions. Lower: Flow accumulation values.

to its downslope neighbour. It also contributes whatever flow it has received itself from its upslope neighbours.

Clearly, for the algorithm to work, we need some way of identifying the downslope neighbour of each pixel. This means there needs to be some way of linking the sorted array of heights back to the original DEM. The simplest is to store the row and column number along with the heights in a 3 column array. The algorithm now becomes:

```
1   Array DEM[0..maxrow,0..maxcol]
2   Array SORTED[0..nrows*ncols,3]
3   Array FLOWS[0..nrows*ncols]
```

```
4  for i=1 to nrows
5    for j=1 to ncols
6       SORTED[id,1]=DEM [i,j]
7       SORTED[id,2]=i
8       SORTED[id,3]=j
9  Sort SORTED array by first column
10     for i=1 to nrows*ncols
11       r=SORTED[i,2]
12       c=SORTED[i,3]
         Add FLOWS [i]+1 to FLOWS[downslope neighbour
13         of DEM[r,c]]
```

Line 1 declares the size of the original DEM. In line 2, a second array is set up to store the heights and row and column ids. Note that since rows and columns are numbered from zero, the number of columns is maxcol+1. Lines 4 to 8 then copy the heights and row and column IDs into SORTED. After the array has been sorted, lines 10 to 13 process the array in height order. The key to this algorithm is line 9, in which the array of heights, row and column numbers are sorted. Sorting a multi-column array like this is no problem, and is what normally required of a sorting algorithm. In Section 5.1, a simple insertion sort was described which would run in $O(n \log n)$ time. The problem in this case is the potential size of the array to be sorted. It is not uncommon to deal with DEMs of 400×400 pixels, giving 160 000 height values. If we then add the same number of row and column ids, we have 480 000 numbers. Assuming each is stored in a 4 byte word, this gives a data size of just under two megabytes. This is likely to present a problem for any sorting algorithm which needs to store the entire set of numbers to be sorted in memory. We saw in Chapter 6 that for a number of reasons, the processing of information which is held in memory is much faster than the processing of information on secondary storage such as a disk drive. However, here we have a situation in which it may simply not be possible to hold the entire array in memory. This is not a problem which is unique to spatial data of course. Commercial databases, such as customer account records, will often be too large to process entirely in memory and yet there will be a need to sort them, to facilitate searching for individual records. Therefore, alternative sorting algorithms have been developed which can deal with these large datasets. One of the commonest is called merge sort, and it is what is called a divide and conquer algorithm. We saw in the case of the insertion sort, and the Delaunay triangulation algorithm, that a useful way to design an algorithm was to try and simplify the problem. In those cases, the method was to deal with the data items one at a time. In divide and conquer algorithms, the problem is simplified by breaking it down into smaller and smaller problems. Eventually, the problem becomes so small that the solution is trivial. The second stage of the algorithm is to combine the solutions to all the simple sub-problems to produce a solution to the original problem.

In the case of merge sort, imagine we have 16 numbers to be sorted. Rather than sort all 16, we split them into two groups of 8 and try and sort each of these. If we repeat this subdivision we will eventually have 8 groups of 2 numbers. Sorting two numbers is a trivial operation. Now we try and start combining our sorted groups of two numbers together. Imagine the numbers written on playing cards. On your left you have a pile of two cards, with the smallest on top and on your right a second pile, also with the smallest on top. To merge the two piles you pick the smaller of the two top cards and place it on a third pile. You then repeat this process with the two cards which are now showing. If you exhaust one pile, then you simply transfer the rest of other pile to the output. This procedure will work no matter how large the piles, as long as each is sorted.

Merge sort is ideal for large files because it is never necessary to process the entire file. In the first part of the algorithm, as much of the file as possible is read into memory, sorted and written out to a new file. The next part of the file is then processed in the same way to produce a second sorted file. When the sorted files are merged, they too can be processed a section at a time, since the records will be dealt with in order.

Merge sort is another algorithm which makes use recursion in its design, as the following code of the algorithm, taken from Cormen *et al.* (1990) shows:

```
1   Procedure MERGE_SORT(A,p,r)
2   if p<r then
3     q=p+r/2
4     MERGE_SORT(A,p,q)
5     MERGE_SORT(A,q+1,r)
6     MERGE(A,p,q,r)
```

A is an array containing elements numbered from p to r. The way this version of the algorithm is written, MERGE_SORT will split the range p to r into two halves and make a recursive call to itself until p is the same as r. When this happens, there will actually only be one element to be sorted and MERGE_SORT exits without doing anything. When this has happened for both calls to MERGE_SORT, the MERGE procedure is called. This assumes that $A[p..q]$ and $A[q+1..r]$ contain two sets of sorted numbers to be merged and returned in $A[p..r]$.

To understand how this can sort anything, it may be helpful to see the sequence of calls which will be made if just 4 numbers are to be sorted. The first call will be to MERGE_SORT(A,1,4) – sort the first four elements of A which are numbered 1 to 4. The sequence will then proceed as follows:

```
 1   MERGE_SORT(A,1,4)
 2     MERGE_SORT(A,1,2)
 3       MERGE_SORT(A,1,1)
 4       MERGE_SORT(A,2,2)
 5       MERGE(A,1,1,2)
 6     MERGE_SORT(A,3,4)
 7       MERGE_SORT(A,3,3)
 8       MERGE_SORT(A,4,4)
 9       MERGE(A,3,3,4)
10     MERGE(A,1,2,4)
```

Remember that as long as there is more than one element to be sorted, MERGE_SORT will call itself twice and then call MERGE. The indentation in the pseudo-code should make it clear that this happens for every call to MERGE_SORT except those on lines 3, 4, 7 and 8.

The MERGE procedure is a little more complicated since it has to process the two sets of numbers it is passed:

```
 1   procedure MERGE(A,p,q,r)
 2   Array MERGED[p..r]
 3   left=p
 4   right=q+1
 5   i=p
 6   while left<=q and right<=r
 7     if A[left]<A[right] then
 8       MERGED[i]=A[left]
 9       left=left+1
10     else
11       MERGED[i]=A[right]
12       right=right+1
13     i=i+1
14   if left>q then
15     Copy A[right..r] to MERGED[i ..r]
16   else
17     Copy A[left..q] to MERGED[i..r]
18   Copy MERGED[p..r] to A[p..r]
```

This rather lengthy diversion makes it clear that it will be possible to sort the array of heights in our DEM flowaccumulation algorithm. If p is the number of pixels in the DEM, the mergesort is an $O(p \log p)$ operation. Processing the sorted list is only $O(p)$ since each pixel is only dealt with once, and so the overall flowaccumulation algorithm is $O(p \log p)$. (Note that this is a change from the previous cases where we have used the size of one side of a raster layer as the size of the problem, in which case the algorithm would be $O(n^2 \log n)$).

FURTHER READING

The book by Wilson and Gallant (2000) gives a comprehensive review of the calculation of many basic terrain parameters using both the gridded DEM and contour-based model. The standard reference for hydrological modelling techniques for gridded DEMs is Jenson and Domingue (1988) although both Hogg *et al.* (1993) and Tarboton (1997), in describing more sophisticated methods also give a good overview of a range of algorithms in this area. Theobald and Goodchild (1990) describe some of the artifacts of TIN-based hydrological modelling, but conclude that these are the result of the way the TINs have been produced rather than any inherent weakness in using this data model for hydrologcical analysis. Van Kreveld (1997) discusses a range of surface algorithms using the TIN, including hydrological modelling, which is explored in more detail by Yu *et al.* (1996). Mark van Kreveld's homepage (http://www.cs.ruu.nl/people/marc/) contains links to online versions of these papers among others.

Skidmore (1989) and Fisher (1993) were among the first to consider the possible effect that different algorithms might have on the results from the analysis of surfaces. Wise (1998, 2000) has considered the effect of different interpolation algorithms in addition to different analytical algorithms on DEM results.

Arge (1997) provides a very good discussion of the additional difficulties in designing algorithms for datasets which are too large to fit into memory.

11 Data structures and algorithms for networks

As described in Chapter 9, although surfaces can be represented using both the vector and raster data model, most current GIS systems tend to use the raster model. This section will consider the example of networks, in which the reverse is true. Because the representation of networks is based on the fundamental data structures used for vector and raster data which have already been described, this chapter has a slightly different structure to the previous ones. The first section tries to indicate, in general terms, why the vector model is most often used for network applications. There are a number of important issues in the sorts of data structures which are used to represent networks, but these are most easily understood in the context of a specific algorithm. The next section therefore describes the algorithm used for finding the shortest routes between points. After a discussion of the effect of different data structures on the operation of this algorithm the chapter finishes with an example of a problem where, not only is there no known, efficient algorithm which guarantees a solution, it is not known whether such an algorithm is possible.

11.1 NETWORKS IN VECTOR AND RASTER

The networks which will be most familiar to many readers are road systems and river channels. Both are sets of lines, which as seen in Section 1.2 can be stored in vector and raster. However, what distinguishes a network from other sets of lines, such as contour lines, is that the lines are connected and this has a crucial bearing on how they are represented in GIS. Figure 11.1 shows a section from an imaginary road network represented using both the vector and raster models.

The most obvious difference is the appearance of the two representations. The raster model gives a poorer representation of the road network, because of the jagged appearance produced by the pixels. This can be improved, but only at the cost of a smaller pixel size, and hence larger files.

Figure 11.1 Imaginary road network represented in vector (left) and raster (right).

We also need to store attributes for each part of the network. These could be stored equally well in either model. With vector, each link would be stored as a line, approximated by a series of points. Each link would also have an ID, which would be used to associate the locational information with the attributes for each link stored in a database. The same approach could be used with raster. With many raster datasets, it is normal practice to store the attribute values directly in the pixels – this is how surface data would be stored for example. However, in the case of networks it is important to maintain the separate identity of each link as an object in its own right. If each pixel simply stores one of the attributes of the link, this separate identity is lost. However, there is no reason why the pixels cannot be used to store a link identifier instead, with the attributes stored in a related database. This would allow the system to answer a range of queries relating to the network such as identifying links which had particular attributes.

However, even with small pixels, and the ability to store the attributes of each link, the raster representation is less well suited to one of the main applications of road network data, which is the production of maps such as tourist route planning maps. For these maps, the link attributes are critical, since these are used to determine the symbolism used for the line. With a vector representation, the data provide the location of the course of the road, and vector graphics can be used to produce lines of different thicknesses, colours etc. With a pixel representation of the route, it is more difficult to produce a range of different symbolisms. It is a very simple matter to change the colours of the pixels which gives a primitive level of cartographic symbolism. However, imagine trying to take part of the road shown in Figure 11.1, and represent it as a dashed line. It would not be possible to do this properly if each pixel is considered in isolation – in order to determine whether a pixel should be ON (i.e. drawn in black and hence part of a dash) or OFF (i.e. not drawn and hence part of the gap between dashes) it is necessary to know how far along the line each pixel is. This information is not stored in the raster data model, and would have to be calculated. In contrast, this sequence information is a fundamental part of the vector model, which stores each link as an ordered set of *XY* pairs.

The raster model is also less well suited to many of the analyses which are carried out using network information, such as finding routes. Planning routes through a network requires two types of data:

1   Information on the 'cost' of travelling along each link.
2   Information on the connections between links.

The simplest measure of how long it will take to travel along each road link will be its length. However, other factors will usually affect how long it takes to travel along a particular piece of road such as, whether there are speed limits and what the volume of traffic is. These can all be generalized as a 'weight' or 'cost' attached to each road link. This information can be handled equally well in vector or raster, since it is just another attribute.

Information about the connections between links is somewhat more difficult to handle in raster than vector. There are a number of ways in which connections between links can be modelled. For instance, a matrix can be drawn up with the links listed along both the rows and columns. Each cell in the matrix can store a 1 if the 2 links are connected, or a 0 if they are not. Alternatively, the cell might contain a number which represents the 'cost' of crossing that junction, since it can sometimes take additional time to cross the junction between two roads. However, a more flexible approach is to use the idea of a node, or a special point which is defined as the junction of two or more links. Nodes are a fundamental part of the link and node data structure used to represent lines in a vector GIS. However, in the raster model, the only spatial entity is the pixel. Even if a special code was used to identify certain pixels as nodes, the raster model has no natural way to represent which links this node connects.

None of these limitations of the raster model are insuperable. If the application needs to estimate travel costs or least cost routes across a whole surface, and not just along a network, then the raster model has considerable advantages over the vector model. However, for applications which need to model flow along a network, the vector model is more straightforward to use. The remainder of the discussion of algorithms and data structures in this chapter will therefore concentrate on the vector model of networks.

11.2   SHORTEST PATH ALGORITHM

One of the most fundamental tasks in network analysis is to find the shortest path between two points. To illustrate the basic algorithm used to solve this problem, we will use the simple network shown in Figure 11.2 – the task is to find the shortest route from point A to point F. Note that the

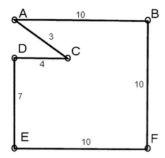

Figure 11.2 Simple network.

figures against each link are simply the distance, in arbitrary units, between the nodes. However, they can represent anything we wish about the link and can be more generally thought of as a weight associated with the link. In a road network, the weight might be related to time taken to travel along the link which means that the weight from A to B might be different from that from B to A. If A was at the top of a hill and B at the bottom, the AB time might be less than the BA time for instance.

This is rather similar to the very first problem which was posed in the introduction – finding a route through a maze. It is very simple for us to see at a glance that there are only two routes from A to F – one via B which has a distance of 20, and one via C, D and E which has a distance of 26. However, the computer of course cannot 'see' the whole network – all it has is information about the nodes and the links between them and will need an algorithm in order to solve the problem. In fact, in a real world application of this problem, such as planning routes for going on holiday or transporting goods, seeing the whole network would not necessarily allow us to arrive at the best solution.

As with most problems, there is a brute force approach, and as with most brute force approaches it will work, but will not be very efficient. In this case, the brute force method is to find all possible routes from A to F, calculate the total distance of each and select the shortest. With the network as it is, this would be quite simple, since there are only two possible routes. However, if we add some extra links, as shown in Figure 11.3, the problem soon grows to unmanageable proportions. Starting at A we have two possible links we can take. Once we get to B, we now have two further links to choose from. One leads to F, which completes one of the routes. The other leads to C, where we have a choice of 3 links which have not been visited yet. This is an example of what is called a combinatorial problem in which the number of possible solutions to a problem grows very quickly as the problem becomes larger. Our original network (Figure 11.2) only had two routes:

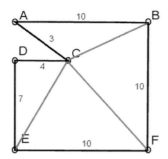

Figure 11.3 Network from Figure 11.2 with extra links added.

Route 1: A, B, F
Route 2: A, C, D, E, F

If we add the link between C and F this adds one more route:

Route 3: A, C, F

However, if we add a link between B and C as well, this adds several more routes, because we now have a greater choice of routes out of C. The new A–F routes are:

Route 4: A, B, C, F – a modification to Route 1 to go via C.
Route 5: A, B, C, D, E, F – a modification of Route 2 to go via B.

As the network size grows, each new link produces a new variation on some proportion of the existing links, so the bigger the network the more potential routes each extra link produces. This type of growth, in which the rate of growth is proportional to the current size of the problem, is called exponential and Figure 11.4 provides some indication of how quickly exponential problems can grow. The column labelled quadratic shows how the problem size grows in the case of $O(n^2)$ complexity, which we have already come across in the context of other algorithms. In the case of exponential complexity, the exponent in the equation is not fixed at 2, as with quadratic growth, but is itself a function of the problem size n – by the time n has reached 7, the exponential problem is already an order of magnitude larger than the quadratic one.

Trying all possible routes is clearly not an option here. An alternative might be to look for the most direct route between A and F – this is what we would do if we were planning a holiday route using a motoring atlas for example. How might this be implemented as an algorithm? Remember that an algorithm has to proceed step by step, considering the choices available at each successive node in the network. Therefore, at each node we would have to pick the link which seemed to take us most directly towards our goal. This could be the link whose direction most closely matched the

n	Quadratic n^2	Exponential x^n
2	4	4
3	9	8
4	16	16
5	25	32
6	36	64
7	49	128
8	64	256
9	81	512
10	100	1024
11	121	2048
12	144	4096
13	169	8192

Figure 11.4 Comparison of quadratic and exponential growth.

overall direction between A and F, or the link whose end point was nearest to F. This might seem like a reasonable approach, and would actually pick the optimum route in Figure 11.3 (A, C, F). However, in the case of Figure 11.2, this approach would generate the route A, C, D, E, F since from node A, the route to C is the one which appears to be leading most directly to F. So this will not work either.

The commonest algorithm in use does not try and find the shortest route forwards from a node. Instead, it selects the link, which when added to the current route will provide the shortest route back to the start point. From the origin all the possible links are examined and the shortest selected – in the example here this is the link to C. From C, D will be picked giving a distance of 9 units from the origin. However, the link to E adds a further 7 units which gives a distance of 16 from the origin. This is further than the direct route from A to B, and so the latter would be selected in preference. Because the shortest route back to the origin is selected at each stage, when the destination is reached, the route found must by definition be the shortest. In fact, this can be proved more rigorously than this, and further reading is given at the end of the chapter for anyone who wants to follow this up.

The algorithm is called the Dijkstra (pronounced Dike-struh) algorithm, after the computer scientist who first described it. It is worth describing in more detail how it is implemented in practice because it illustrates some important points about how the network is stored as a data structure. This description is based on that given by Worboys (1997) – the reading list provides details of other alternative accounts and even of some animations of the algorithm on the World Wide Web (W W W). For the moment let us assume that for each node, we can work out which links start at that node and how long they are. We set up three arrays of information for each node:

- *Distance*: The total distance from this node back to the origin. This is set to zero for the start node and infinity for all other nodes.
- *Parent*: The next node back along the route to the origin. This is set to blank for all nodes.
- *Included*: Whether or not this node has ever been included as part of the shortest route. This is set to NO for all nodes.

At the start of the algorithm, the arrays are as shown in the table below

	Distance	*Parent*	*Included*
A	0	–	NO
B	∞	–	NO
C	∞	–	NO
D	∞	–	NO
E	∞	–	NO
F	∞	–	NO

The algorithm is described below. The notation dist(N) indicates the current value of the distance column for node N. The notation d(NM) indicates the distance between nodes N and M.

- Look at all nodes which have not been included so far (whether or not they have a link with the current position) and choose the one with the shortest current distance back to the start.
- Mark it as YES in the included array. Let us call it N. The distance from this back to the start is dist(N).
- Find the nodes which are connected to N and which are still marked NO in included. For each of these M nodes, the distance between it and N is d(NM).

```
1  for each node
2     if (dist(N)+d(NM))<dist(M) then
3        dist(M)=dist(N)+d(NM)
4        parent(M)=N
```

In other words, if there is a shorter route back to the start from node M via node N, then the arrays are updated to record this fact.

In the first iteration, the node with the smallest distance value is A, since it has a distance of 0. This is picked and its included value set to YES. Two nodes are connected to A – B and C. Their current distances back to the start are infinity. In both cases then, the route back to A is shorter than this, so their distance values are updated, and their parent node is set to A. This produces the following set of arrays

	Distance	Parent	Included
A	0	–	YES
B	10	A	NO
C	5	A	NO
D	∞	–	NO
E	∞	–	NO
F	∞	–	NO

Table after one iteration

In the second iteration, nodes B to F are considered and C found to have the shortest current distance back to the start. The included column is now updated to YES for C, so it will no longer be considered in the algorithm. It only has one neighbour, D, which is 4 units away from C. Adding this distance, to the distance from C back to the start (5) gives a distance from D back to the start of 9 units. This is less than the current distance for D, which is still infinity, so the distance column for D is updated with a value of 9 and its parent is set to C.

	Distance	Parent	Included
A	0	–	YES
B	10	A	NO
C	5	A	YES
D	9	C	NO
E	∞	–	NO
F	∞	–	NO

Table after two iterations

In the next iteration, D is the node with the shortest current route back to the start. It only has one neighbour, E, and so after this iteration, the table will look like this:

	Distance	Parent	Included
A	0	–	YES
B	10	A	NO
C	5	A	YES
D	9	C	YES
E	16	D	NO
F	∞	–	NO

Table after three iterations

From E, there is a direct route to F, but we know that this would not be the shortest. What the algorithm does is to pick node B next, since this has not so far been included on the route, but now has the shortest route back to the start. B only has one neighbour which has not yet been included, F. The distance BF, plus the distance back to the start from B are added together and put into the distance column for F. The updated table now looks like this:

	Distance	Parent	Included
A	0	–	YES
B	10	A	YES
C	5	A	YES
D	9	C	YES
E	16	D	NO
F	20	B	NO

Table after four iterations

The next node to be considered will be E, since of the two nodes which have not had their included value set to YES this has the shorter distance back to the origin. The distance from E to the start is 16. The distance from E to its only non-included neighbour is 10. Since the sum of these two is greater than the current distance back from F to the start, no change is made to the distance column for F. After the fifth iteration the table looks as follows:

	Distance	Parent	Included
A	0	–	YES
B	10	A	YES
C	5	A	YES
D	9	C	YES
E	16	D	YES
F	20	B	NO

Table after four iterations

On the next iteration, the only node left to be considered is the destination, and this is where the algorithm terminates. The final table shows that the shortest path from F back to A is 20 units long. The actual route can be determined by tracing back through the parents – from F back to B and hence to A. Another interesting feature of the Dijkstra algorithm is that the final table actually contains the shortest distance and route from every node back to the start point.

The efficiency of the algorithm depends on the number of nodes (n) and the number of links (l) in the network. On each iteration of the algorithm, one node is added to the list of those INCLUDED. Since this happens only once for each node, there are $O(n)$ iterations of this main loop. Each link has a distance associated with it. This is considered just once – when one of the nodes which it joins is added to INCLUDED. When the node at the other end is later added to INCLUDED, the distance is not considered again.

Hence, each link is considered once making the total complexity $O(n + l)$. However, there is one further crucial element to consider. On each of the n iterations of the main loop, the node with the shortest distance back to the origin, which has not already been considered, has to be selected. If the nodes are stored in a simple array, as shown in the tables, the only way of finding the one with the shortest distance is to search the whole array which is an $O(n)$ operation. Since this is done on each of the n loops, this makes the total complexity of the algorithm $O(n^2 + 1)$, which is $O(n^2)$.

So far we have simply assumed that we can derive all the information we need to know about the network in order to operate the algorithm. In the next section, we will consider what data structures we will actually need in order to do this, and from this we will see how we can improve the efficiency of this algorithm from $O(n^2)$ to $O(n \log n)$.

11.3   DATA STRUCTURES FOR NETWORK DATA

We have already seen in Section 2.3 the storage of lines and the connections between them in describing the basic vector data structures. Figure 11.5 shows the same basic network but with each link represented by a number, rather than its weight.

Using the link and node data structure, based on the topological structure described in Chapter 2 we could represent this network as shown in Figure 11.6.

The question is whether we can use this table to fill in the three arrays we need for the Dijkstra algorithm. The algorithm consists of three steps, which can be simplified to:

1   From the nodes which have not been included, pick the one with the shortest value in the distance array.
2   Mark this as included and note its current value in the distance array.

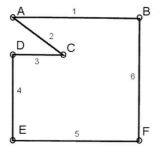

Figure 11.5 Simple network with link identifiers.

Link ID	From node	To node	Weight
1	A	B	10
2	A	C	5
3	C	D	4
4	D	E	7
5	E	F	10
6	B	F	10

Figure 11.6 Link and node representation of simple network.

3 Find the nodes which are connected to this one. For each connection, find the weight assigned to the link. See if this provides a shorter route back to the start than the current one.

The first two steps can be handled entirely from the distance and included arrays, although we will see later in this chapter that this is not the most efficient way to handle these steps. First, let us consider the third step and start by considering the situation when node C has been picked as the current node. We now need to find out:

1 Which nodes have a direct link to C?
2 Which of them have already been included?
3 Of those which have not, what is the weight of the link from node C?

All of this information can be supplied using the table in Figure 11.6. To answer the first query, we would need to search the table and extract all rows which had C as either the start or end node. In the context of the current operation, we do not care whether a link goes from A to C or from C to A. This would pull out the records for links 2 and 3. Now we note which node is at the other end from C in each case. We check in the included table to see which have already been considered. This will leave only link 3, since node A has already been dealt with. From the record for link 3, we can read the weight which provides the answer to the third query.

This seems fine in this simple case. However, consider what would happen if instead of six links we were dealing with 600, or 6000 or even 60 000. The first of our three queries above now becomes a matter of selecting two links from a total of 60 000. If we simply start at the first row and search a row at a time, this is going to be very inefficient, especially when we consider that this is only part of an algorithm which will be

Node	Links
A	1,2
B	1,6
C	2,3
D	3,4
E	4,5
F	5,6

Figure 11.7 Node table for network.

repeating this operation at least once for every node in the network. It is clearly worth trying to find a quicker way to answer this query.

One option is to create a second table, which lists the links which are connected to each node (Figure 11.7).

This table can be created in $O(n)$ time from the table in Figure 11.6. Now our query becomes much quicker because we can go directly to the correct entry in the node table, since we know which node we are looking for. The entry will tell us which links we need to look at, so we can extract these directly from the link table – there is no need to search every link record.

In fact, the basic table we used for the Dijkstra algorithm also had a record for each node, so rather than have an extra table it makes sense to add the information about the links connected to each node to this table. The problem with the table in Figure 11.7 is that different nodes may have different number of links connected to them which is messy to deal with. One solution to this is to modify the link table in Figure 11.5 so that it is what is called a linked list, as shown in Figure 11.8.

The node table has been modified to provide a pointer to one of the links which starts at that node, and one of the ones which ends there. In the link table, if more than one link starts at the same node, the record for the first link contains a pointer to the next link. If there are no further links, then the pointer contains −1. To find out which nodes neighbour C, we can use these two tables to search through two lists in the link table – the list of links which start at C and the list of those which terminate at C.

This data structure will speed up the operation of the Dijkstra algorithm, by making the search for neighbouring nodes faster than if we searched the

Link ID	From node	To node	Weight	Next link from start	Next link to end
1	A	B	10	2	−1
2	A	C	5	−1	−1
3	C	D	4	−1	−1
4	D	E	7	−1	−1
5	E	F	10	−1	6
6	B	F	10	−1	−1

Node	Distance	Parent	Included	First link out	First link in
A	0	-	NO	1	−1
B	∞	-	NO	6	1
C	∞	-	NO	3	2
D	∞	-	NO	4	3
E	∞	-	NO	5	4
F	∞	-	NO	−1	5

Figure 11.8 Storage of network information using separate node and link tables.

original table of information for the links. However, it will not improve the overall efficiency of the algorithm. As seen at the end of Section 11.2 this is still $O(n^2)$ because we process each node once, and in each case we need to find the node with the shortest distance back to the origin which has not so far been considered. If we store the distance in a simple array, as seen in Section 11.2, this second step will be an $O(n)$ operation. However, if we can sort the distances into order, we will be able to find the shortest distance more quickly. In fact, if the distances are sorted into order, the shortest is the first in the list so retrieving it becomes $O(1)$. However, this is not simply a matter of taking an array and sorting it, which we know can be done in $O(n \log n)$ time. The values in the array will change, as the distances are updated, and the elements will be removed from consideration when they are added to the INCLUDED list. This means we need some way of updating our sorted list as changes are made, without having to completely resort it. A data structure which can do this is called a priority queue, and a common form of priority queue is the binary heap.

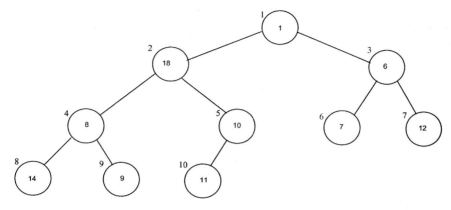

Figure 11.9 Heap. The figures outside the nodes are the node identifiers (1–10). The figures inside are the values held in the node. Node 2 violates the heap property since its value is greater than those of its children.

This is a tree data structure (like the quadtree which was described in Section 6.2) with two properties:

- Each node in the tree has a maximum of two children (hence the binary in the name).
- The children are both equal to or larger than their parent.

The second property means that the root node of the heap contains the smallest value, which is why it is so useful as a priority queue. What is more, all the operations on the tree, such as building it and maintaining the two heap properties can be done in O(log *n*) time.

To illustrate how the binary heap works, let us start with the example shown in Figure 11.9. This shows a heap containing 10 numbers. The tree is what is called a full tree. This means that starting from the root, each level in the tree is completely filled by giving each node two children. This may not be possible in the lowest level of the tree, which is filled from left to right as far as possible. This means that if any node apart from the last one is removed, the other nodes have to be moved in order to fill the gap.

All the nodes in this case have the second property of being smaller than their children, except node 2. In order to restore the heap to the proper form, we use a procedure called HEAPIFY which has the following structure.

```
1  Heapify (startnode)
2  Find which is smallest from startnode, left-child,
     right-child.
3  If startnode not smallest
```

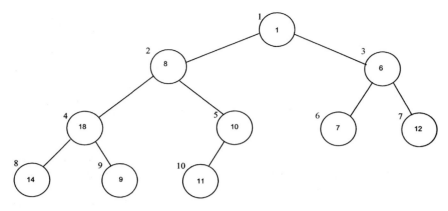

Figure 11.10 Heap after the swap between nodes 2 and 4.

```
4   Swap smallest with startnode
5   Heapify(smallest)
```

This will start at node 2 and examine the values in nodes 2, 4 and 5. The smallest of these is node 4 (value 8) which is not the current startnode. The values at nodes 2 and 4 are swapped to give the situation shown in Figure 11.10. This swap means that startnode is now smaller than both its children, restoring the heap property. However, the child which was swapped may now be larger than its children (which is the case in this example) and so heapify is called recursively starting at the swapped child node.

In order to implement this procedure, we need to be able to identify the left and right child of any given node in the tree. This is made very simple by the fact we are dealing with a full, binary tree:

```
left(i)=i*2
right(i)=i*2+1
```

We have already seen in Chapters 7 and 8 that mulitiplication by 2 is very efficient on computers, since it can be accomplished simply by shifting the bits in a number one position to the left. Finding the parent of a node is equally easy:

```
parent(i)=i/2
```

Note that this uses integer division, in which the remainder is ignored. This means that $5/2 = 2$ and not 2.5. Again division by 2 is achieved by shifting the bits in the number to the right.

Node	1	2	3	4	5	6	7	8	9	10
Value	*1*	*18*	*6*	*8*	*10*	*7*	*12*	*14*	*9*	*11*

Figure 11.11 Storage of the binary heap from Figure 11.9 in an array. The upper row indicates the position of the values in the array.

This simple relationship between the identifiers of parent and child nodes means that a binary heap can be stored in a simple one-dimensional array as shown in Figure 11.11.

The array itself is shown in the lower row of Figure 11.11. The upper row is simply to help you to confirm for yourself, that the operations for identifying the parents and children of nodes work properly.

We can now present a more detailed version of Heapify, taken from Cormen *et al.* (1990).

```
1   HEAPIFY(A,i)
2   /* Find ID of child nodes
3   l=left(i)
4   r=right(i)
5   /* Find ID of which has the smallest value
6   if l<=heap_size and A[l]<A[i]
7     then smallest=l
8     else smallest=I
9   if r<=heap_size and A[r]<A[smallest]
10     smallest=r
11  /* See if current node is smallest
12  /* If not swap its value with the smallest
13  /* and call HEAPIFY starting at swapped child
14  if i!=smallest
15    temp=A[i]
16    A[i]=A[smallest]
17    A[smallest]=temp
18    HEAPIFY(A,smallest)
```

If HEAPIFY is run starting from the root node, the maximum number of iterations will be determined by the number of levels in the tree. Since the tree is binary, the maximum number of levels is $\log n$, hence HEAPIFY will run in $O(\log n)$ time.

This is important, because every time we extract the minimum distance from the heap we will leave the root node empty. How do we restore the heap to its proper form? Consider the example in Figure 11.12. If the value

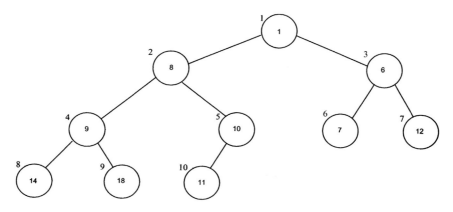

Figure 11.12 Heap after swap between nodes 4 and 9. All nodes now satisfy both the heap properties.

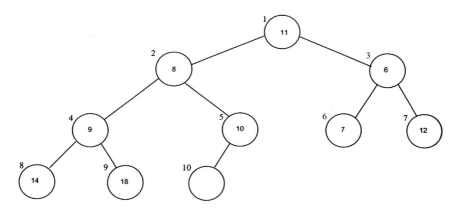

Figure 11.13 Heap after the removal of the root node, and its replacement by the value from node 10. Node 10 is empty and can be removed from the tree.

of 1 is removed from the root node, we have 9 nodes instead of 10. We know that when we have finished updating the tree, it is node 10 which will be empty, since the tree is always filed from top to bottom and from left to right. Therefore we take the value from node 10 and place it in the empty root node giving the situation shown in Figure 11.13. The new root node value will probably be larger than its children (as it is in this case) but we already have a HEAPIFY procedure to deal with this. In this example, two swaps will restore the heap to its proper form as shown in Figure 11.14.

Adding an extra node is rather similar. We add a new node at the bottom right hand side of the tree, and then run a version of HEAPIFY which checks the tree rooted at the parent of the start node. As the Dijkstra algorithm

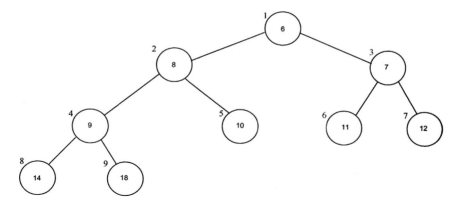

Figure 11.14 Heap from Figure 11.13 after two swaps – nodes 1 and 3, followed by nodes 3 and 6.

proceeds, the distances associated with nodes in the route network will also change, which will require further calls to HEAPIFY. Since all these operations run $O(\log n)$ time, we can see that the Dijkstra algorithm using a binary heap as a priority queue will run in $O(n \log n)$ time.

There are two slight complications in the use of a binary heap. First, the examples so far have only shown distances stored in the heap. However, we would need to store the actual ID of the node in the network along with its current distance value. This could be done by using two positions in the heap array for every node in the heap, and altering the left, right and parent functions accordingly. However, we also need to be able to find the position of any given network node in the heap, so that we can update its distance value. The heap is not an efficient structure for this type of query and this would necessitate a linear search through the heap array. To avoid this, we would add an additional pointer to our node table, which pointed to the position of each node's record in the heap array. This would have to be updated every time we modified the heap of course, but this is an $O(1)$ operation, so has no effect on the overall complexity of our algorithm.

This section has further developed the idea that the design of data structures is just as important as the design of algorithms in the development of efficient computer programs. We will end our consideration of networks with an example of a problem for which no efficient algorithms are known to exist, and where alternative strategies have to be adopted.

11.4 THE TRAVELLING SALESMAN PROBLEM

In Chapter 1, the design of computer algorithms was illustrated with the example of finding a route through a maze. A solution which was suggested was to place your left hand on the left hand wall of the maze and keep it there. For some mazes this will work but there are some mazes for which it will not. Strictly speaking therefore this cannot be considered an algorithm since it does not guarantee a solution. Instead it is what is called in the computing literature a heuristic – something which does not guarantee a solution, but which can still help lead to the solution of a problem none the less.

In the shortest path problem, we have seen that there exists an algorithm which can be mathematically shown to produce the shortest route between two points. But in some situations it may be enough to find a route which is not too long rather than the one which is absolutely the best. For instance, imagine arriving in a new city and trying to find the way to your hotel using a street map. In this case, you would probably plan a route by setting off in the right general direction, and taking whichever roads seemed to lead most directly to the hotel. This is also a heuristic. It will not give the best solution. On the other hand, in most cases, it will produce a solution which is good enough.

There are some problems for which no efficient algorithms are known and in which heuristics are all we have. A classic example is the travelling salesman problem (TSP) which is deceptively simple when stated. Given a set of points in space, plan the shortest possible route which visits each point exactly once and returns to the start. Despite a great deal of effort there is still no known algorithm which provides an efficient solution to this problem. Good solutions to the problem are of real practical application – any organization which needs to deliver goods or visit clients has an interest in planning itineraries which are as short as possible. Intriguingly, there is a very similar problem, for which algorithms do exist. This is called the Minimum Spanning Tree (MST) problem, and the only difference is that the route only needs to connect all the points using the shortest path – it does not need to return to the start. More formally stated the MST problem is to find the shortest path which allows someone to travel from any point to any other point. To see why this is different from the TSP consider Figure 11.15. The route on the left is the TSP solution for these four points. However, to solve the MST problem, the direct link between 1 and 4 is not needed since to get from 1 to 4 it is possible to go via points 2 and 3.

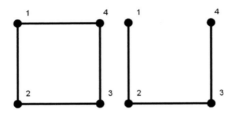

Figure 11.15 Solution to the TSP (left) and MST (right) for four points.

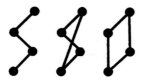

Figure 11.16 Illustration of why the MST does not form an optimal solution to the TSP.

The route on the right is the MST for these four points – all the points are connected and the total length of the path is as small as possible.

It might seem that since the difference between the problems is so slight, that the MST should form a good solution to the TSP. It is true that it often forms a good approximation, but it is not hard to show that it does not form a solution to the problem. Figure 11.16 shows four points, and on the left the minimum spanning tree, which is 15 units long. The only way to convert this to a closed route is to join the first and last points, as shown in the centre of Figure 11.16, an extra distance of just over 12 units. The TSP route for these points is shown on the right, and is only 26 units in length.

Consider the points in Figure 11.17. The distances between points are 3 units in the north-south direction, 4 units in the west-east direction, and 5 units diagonally. No roads have been shown in the diagram, and it is assumed that it is possible to travel between any pair of points. In the real world this would not be true, but a general algorithm could easily be modified to deal with this. For the purposes of illustrating the features of the TSP, it is easier to ignore this level of practical detail.

Figure 11.18 shows some alternative routes which visit each location just once, and their total travel distances.

The best route is clearly the one which makes greatest use of the shorter north-south roads, the worst is the one which uses the long diagonal roads most. An obvious approach for finding a solution to the TSP is therefore to start at a point, and from there move to the closest, unvisited point. If we

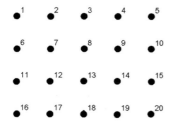

Figure 11.17 Street network for TSP.

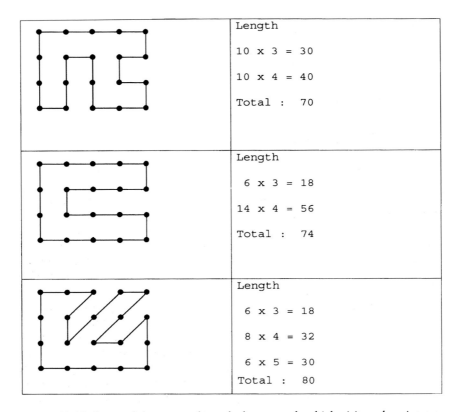

	Length
	10 x 3 = 30
	10 x 4 = 40
	Total : 70
	Length
	6 x 3 = 18
	14 x 4 = 56
	Total : 74
	Length
	6 x 3 = 18
	8 x 4 = 32
	6 x 5 = 30
	Total : 80

Figure 11.18 Some of the routes through the network which visit each point once.

begin at point 1, this will take us down to point 16, along to 17, and then up to 2, along to 3 and so on. Eventually, we will reach point 20 and will hit a problem. What we have actually done is trace the MST and we have exactly the problem shown in Figure 11.16. The only way to close the tour is by joining the furthermost two points together and this gives a total distance of just over 83 for the route. In fact, the only known algorithm which will solve this problem is the brute force algorithm – list all possible routes and pick the shortest. It should be apparent, that like the shortest route problem, this is a combinatorial problem, and as the number of points to be visited grows, the number of routes to be calculated soon outstrips the power of existing computers.

We will come back to the issue of algorithms for this problem in a moment. First, let us consider some heuristics. The design of heuristics is rather similar to the design of optimal algorithms and some of the strategies used for designing good algorithms will lead to good heuristics. A strategy which is commonly used is solve the problem with a small subset of the

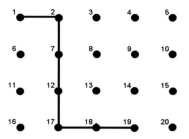

Figure 11.19 Incremental heuristic after 3 points.

data, and then add in the additional points. This was the format of the incremental sort algorithm (Section 5.1) and also of the algorithm for creating a Delaunay triangulation of a set of points (Section 8.3). In the case of the travelling salesman one approach would be to take two of the points and find the shortest route between them. Then we take another point at random, and find the shortest detour from the existing route which will include the new point.

Assume our first points are 1 and 7. There are two equally short routes, so we pick one at random. The next point is 19. This is easy and our route so far is shown in Figure 11.19. Note that we have had to pass through several other points on our way, so we add these to the current route. The next two points are 14 and 16. These are close by the current route, so might not be expected to cause problems but as we shall see they do. Consider point 14. One way to generate a detour is to replace one link in our current route with a route which visits the new point. The obvious link to replace is the one between the two closest points – 18 and 19. The shortest route between 18, 19 and 14 includes one of the diagonal links as shown in Figure 11.20. With point 16 the two closest points are 12 and 17 and replacing their link with a detour to 16 adds another diagonal to the route.

The sequence of points in this case was chosen to illustrate some of the problems with this heuristic. However in general, any system which picks

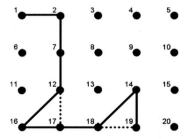

Figure 11.20 Incremental heuristic after 5 points.

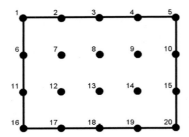

Figure 11.21 Incremental approach starting with convex hull.

points at random is sensitive to the particular sequence of points. An alternative approach is to try and consider what the characteristics of a good route would be. The ideal is a perfectly circular route, since this would contain no deviations at all. This is impracticable of course, but perhaps rather than starting with points picked at random we could start with the closest we can come to a circular route – one which joins all the outer points as shown in Figure 11.21. Now rather than picking the remaining points at random, let us add the point which causes the least deviation from this route. In fact, all the remaining points can be added by creating a deviation from two of the points on the outer route, giving a final route as shown in Figure 11.22. The length of this route is 74, which is about halfway between the optimum and the worst routes in Figure 11.18.

The problem with the heuristics so far is that each additional point has been considered in isolation. For instance, if we had added points 7, 8, 12 and 13 at the same time as a diversion between points 2 and 3 this would have produced a shorter route still. However, it is much easier to produce heuristics which consider one point at a time, than ones which consider several points simultaneously. One of the strengths of algorithms, such as the incremental algorithm for creating a Delaunay triangulation is that it allows the situation around each individual point to be dealt with but in a way which guarantees an overall optimum solution. The reason that this is

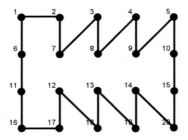

Figure 11.22 Route after adding all points to outer route shown in Figure 11.21.

possible in this case is that the characteristics of the optimum solution are known. It is known that the Delaunay triangulation is an optimal triangulation of a set of points, that there is a unique Delaunay triangulation of any given set of points and what many of its properties will be. The same is not true of the TSP which is part of the reason that there is no algorithm which is guaranteed to produce an optimal solution.

The question of whether such an algorithm can ever be produced is one of the great unanswered questions of computational analysis. In one sense, of course an algorithm does exist to solve the TSP – the brute force algorithm which lists all the possible solutions and pick the best. However, like the brute force approach to the shortest route problem this has exponential complexity and is impracticable for any but the smallest networks. The question is therefore not whether an algorithm is possible, but whether an algorithm which has a reasonable degree of complexity is possible. Reasonable is normally taken to mean polynomial complexity i.e. $O(n^2)$ or $O(n^3)$. All algorithms which have polynomial complexity or better are called class P algorithms. Any algorithm which can be run in polynomial time can also be checked in polynomial time. For instance, given the output of a sort routine, it is possible to check that the algorithm has indeed sorted the numbers in $O(n)$ time. Each number is checked in turn and if one is found to be out of sequence the check fails.

Interestingly, although it may not be possible to solve a problem in polynomial time, it is often possible to check a proposed solution in polynomial time. Such algorithms are called NP, which stands for non-deterministic polynomial. Non-deterministic means that a solution to the problem can be found (the problem is not completely intractable) but that this solution may involve making an educated guess and then seeing if that has in fact worked. Polynomial means that this check can be done in polynomial time. The TSP is an example of an NP problem – heuristics exist which will generate solutions, and it is possible to test whether these are in fact optimal.

It is known that all problems which belong to the class P are also NP. However, it is not known whether NP problems also belong to the class P. Why does this matter? If it could be proved that NP problems cannot be solved in polynomial time, then there would be no need to carry on trying to develop algorithms for their solution – we would know that heuristic approaches are the best we are ever going to get. Conversely, if it could be proved that NP problems did belong to class P, this would make it worth trying to find an algorithm for a problem such as the TSP.

But is it possible to prove that a problem like the TSP can be solved in polynomial time without actually doing it? Surprising as it may seem, the answer is yes. The reason is that all computing problems, no matter how simple or complex, can be analysed and compared in a way which does not depend on their exact details. In the 1950s, the English mathematician, Alan Turing, proposed the idea of what has come to be known as a Turing machine. This is a 'computer' which consists of an infinite length of tape which can be read by a small, moveable head. The tape contains sequences of symbols, and depending on the symbol which is read, and the 'state' in which the reading head is in at the time, the head will perform one of a series of actions – move left, move right, write a new symbol. The remarkable thing is that it has been shown that any form of computation, from adding two numbers to modelling the earth's climate, could in theory be programmed on a Turing machine. The key to understanding this, is to understand that any problem can be broken down to a series of smaller problems. For instance, in Section 6.1 we saw that any mathematical expression could be broken down to a series of operations which involved two quantities at most. Hence, a computer which can add, subtract, divide and multiply two numbers can carry out an enormous range of calculations. Multiplication itself is simply addition repeated several times. Three times three is actually $3 + 3 + 3$. So in order to perform multiplication, the computer only needs to be able to add and to count.

The importance of the Turing machine is that if all problems can be run using the same set of simple operations, then they must all be in some way comparable with each other. If there were a set of computations which could not be run on a Turing machine, then these would form a completely different class from those that could. (As an interesting aside, many people believe that the human brain is capable of carrying out operations which cannot be programmed on a Turing machine, and that this means that current computers will never be programmed to reproduce human intelligence.) Because of this comparability between computer programs, it is possible to analyse a particular algorithm and assign it to a class such as P and NP. It is also possible to show that certain problems, such as the TSP, can be converted to be the same as a different problem. To take a trivial example, it is easy to see that if we want to find out whether a point falls inside a rectangle, we do not necessarily need a special algorithm – we can use one which will determine if a point falls within an area of any arbitrary shape. This will be less efficient of course, but that is not the issue here. Once we have an algorithm for problem A we can use this to prove that an algorithm must exist for any other problem B, which can be converted to problem A. This conversion will not necessarily be a good solution in itself, but the proof that an algorithm must exist, means it is worth the effort searching for an efficient solution to problem B.

Returning to the TSP, this belongs to a special set of NP problems called NP-complete. These are NP problems which can be shown to all be equivalent to one another. If it can be shown that an efficient algorithm exists to solve one of them, this would prove that efficient algorithms exist for all of them. Conversely, if it could be proved that no efficient algorithm exists for one of them, this proves that none exists for any of them. Proving which of these is true remains one of the great unsolved problems of computational complexity. Until it is solved, programmers will continue to work on better heuristics to provide practical solutions to problems such as the TSP.

FURTHER READING

Gardiner (1982) provides a nice discussion on the importance of structuring network data, in the context of handling information on river networks. Haggett and Chorley (1969), although writing in a largely pre-computer age, cover many of the issues which arise in trying to solve problems using networks.

The Dijkstra algorithm for shortest route is described by both Jones (1997) and Worboys (1995). Interactive demonstrations of the operation of the algorithm can be found online at http://swww.ee.uwa.edu.au/~plsd210/ds/dijkstra.html and http://www.cs.uwa.edu.au/undergraduate/courses/230.300/readings/graphapplet/graph.html

The Travelling Salesman problem has intrigued people for a long time, and it has spawned a wealth of web sites. The maths department at Princeton University has an excellent web site (http://www.math.princeton.edu/tsp/index.html) which includes a pictorial history of the solutions to some notable problems, such as an optimal route between 13 509 US cities. Both Gunno Tornberg (http://w1.859.telia.com/~u85905224/tsp/TSP.htm) and Stephan Mertens (http://itp.nat.uni-magdeburg.de/~mertens/TSP/index.html) have pages which include animations of different heuristics for solving this problem. Finally, the TSPBIB home page (http://www.densis.fee.unicamp.br/~moscato/TSPBIB_home.html) aims to 'be a comprehensive listing of papers, source code, preprints, technical reports, etc., available on the Internet about the Travelling Salesman Problem (TSP) and some associated problems'.

A range of other network algorithms, including the optimum location of centres to provide services to sites on a network (the location allocation problem) are covered in one of the units in the second version of the NCGIA core curriculum (http://www.geog.ubc.ca/courses/klink/gis.notes/ncgia/u58.html).

Conclusion

This book has tried to provide an introduction to the key issues involved in the design of computer software for handling spatial data. The approach which has been taken has been to select a few key data structures and algorithms and describe these in some detail. Inevitably, this approach means that many important topics have not been covered. For instance, Section 2.3 described the georelational method of storing vector data, in which the attributes are held in a relational database, while the locational data is handled separately using purpose written software. Many systems in use today still use this architecture. However, some of the problems which meant that this separation of the two elements of the data was sensible have now been solved. Database vendors have produced databases which are capable of storing spatial data and query languages which include simple spatial constructs. The GIS vendors have realized the advantages to be gained from using standard database technology to handle both locational and attribute data and have begun to market systems which use this integrated approach. Alternatives to the relational database are also being developed, and future GIS software may well be based on object-oriented databases for instance.

However, the advantage of the approach which has been taken is that it has been possible to explain some important ideas and concepts fully. Whatever new developments take place in GIS software, you can be sure that they will be underpinned by the design of efficient data structures and algorithms, even if the resulting systems appear quite different from the current generation of GIS. I very much hope that having read this book you will not only be inspired to keep up with some of the future developments in the field of GIS, but that you will be better prepared to follow them.

Glossary

This glossary contains a set of informal definitions of some of the terms introduced in this book. Common GIS terms, such as vector and raster, which are described in standard GIS textbooks are not included.

1-cell   One of the names for the lines connecting nodes in the **link and node** data structure for representing vector data. Others are **chain, edge, link, segment** and **arc**.

Address   Number which uniquely identifies one unit of storage in computer memory. Computer circuitry is designed so that information can be moved between the CPU and all addresses equally quickly.

Algorithm   A procedure or set of steps to provide a solution to a problem. It should be possible to prove that the algorithm will always provide a solution (see **heuristic**).

Application domain model   A representation of some element or elements of the real world in a form which is comprehensible to experts in a particular application area. A map may be regarded as an application domain model.

Arc   One of the names for the lines connecting nodes in the **link and node** data structure for representing vector data. Others are **chain, edge, 1-cell, segment** and **link**.

Array   A general purpose computer data structure which can store a list of data elements. Individual elements can be accessed extremely efficiently in memory.

Big O notation   Informally, if an algorithm has $O(n^2)$ complexity, then as the number of objects to be handled (n) increases, the number of operations to be performed (and hence the approximate time taken) increases as the square of n. More formally if $f(n) = O(g(n))$, $g(n)$ is always less than $f(n)$ times a constant value for values of n above a second constant value. So $n^2 + 4$ is $O(n^2)$ because $n^2 + 4$ is always less than $2n^2$ for values of n above 4.

Binary search tree   A data structure used for storing and searching sorted data. Each subtree has the property that all nodes to the left of the root are smaller than or equal to the root, and all nodes to the right are larger than or equal to the root (see **tree**).

Bit interleaving   A technique for combining the X and Y coordinates of a location into a single spatial key. The coordinates are represented as binary integers. A new value is created by taking 1 bit at a time from the X and Y coordinates in turn (see **Morton code**).

Bit shifting   Operation which moves the **bits** in a binary **word** or **byte** to the left or right. If the bits represent a binary integer, a shift of one bit is equivalent to multiplying or dividing by 2.

Bit   Binary digit (0 or 1 in base 2 arithmetic). The smallest unit of storage within a computer file or in memory.

Branch   A **node** in a **tree** which has further nodes beneath it.

Byte   Unit of storage consisting of 8 bits.

Centroid   A point which is used to label an area. It must lie within the area, and should ideally be located near the centre.

Chain   One of the names for the lines connecting nodes in the **link and node** data structure for representing vector data. Others are **link**, **edge**, **1-cell**, **segment** and **arc**.

Complexity   A measure of the efficiency of an algorithm. It is expressed as a function which describes how the time or storage needed to solve a problem, increases as the size of the problem increases. Most commonly described using **big-O notation,** which describes the worst case behaviour of the algorithm.

Computer precision   The number of significant digits which can be reliably stored and used in calculations in a computer **word.**

Conceptual computational model   A representation of an **application domain model** in a form which is potentially representable on a computer, but which is not specific to any particular language or package. The vector and raster models may both be regarded as conceptual domain models.

CPU   Central Processing Unit of a computer. Capable of performing operations on data held in the memory of the computer. As well as standard arithmetic operations, such as addition, usually able to perform operations such as bit shifting. Often contains a small number of **registers.**

Data model   A representation of a selection of objects or phenomena from the real world in a form which can potentially be stored on a computer. The data model itself contains no details of the computer implementation and is purely a conceptual construct. In non-spatial databases, the relational model is commonly used. The vector and raster models are commonly used for spatial data.

Data structure   Describes any structure used to organize data in the memory of a computer. Common structures include the **array** and various forms of **tree**. Specialized structures have been developed for the storage of spatial data.

DIME   Dual Independent Map Encoding. A **data structure** for the storage of address-based street information developed by the US Bureau of the census.

Double precision   The storage of floating point numbers in a 8 **byte** (64 bit) **word**. This means that calculations only have a **precision** of approximately 15 significant digits.

Edge   (1) In graph theory, the connection between nodes in a **graph**. (2) In GIS one of the names for the lines connecting nodes in the **link and node** data structure for representing vector data. Others are chain, link, 1-cell, segment and arc.

Face   In graph theory, the area defined by the edges of a **graph**. If a graph has no cycles (closed loops) there is only 1 face.

Floating point   Name for the method used to store numbers with a fractional part in the computer. The digits are stored in the **mantissa**, while the position of the decimal point is stored in the **exponent**.

Geometrical data   Data which stores information about the location of vector objects. Sometimes loosely referred to as geometry (see **Topological data**).

Georelational   A name applied to GIS systems in which the locational (or **geographical**) data is stored separately from the attribute data, which is held in a relational database.

Graph   A representation of the set of connections between items. The items are represented as points, called **nodes**, and the connections between them as lines, called **edges**. The areas which are defined by the network of edges are called **faces**. A map of a road network can be regarded as a graph. **Tree** data structures can also be represented as graphs.

Heuristic   A procedure or set of steps which are generally found to help find a solution to a problem. In contrast to an **algorithm,** a heuristic does not have to solve the entire problem, and does not guarantee to provide good solutions in all cases.

Hexadecimal   Storage of numbers to base 16. Letters A to F are used to represent the digits above 9.

Interpolation   The estimation of an unknown value of a variable or parameter from known values. Strictly, the known values must lie on either side of the unknown value. If the unknown value exceeds the known ones, or lies outside the spatial extent of the known ones, the process is known as extrapolation.

k-d tree   A data structure used for point data. The area is subdivided according to the X coordinate of points at even levels in the tree, and the Y coordinate of points at odd levels. The structure allows for efficient spatial searches of points (see **tree**).

Leaf   A **node** in a **tree** which does not have any further nodes beneath it.

Line segment   Straight line connecting two points, forming part of the digitized representation of a line.

Link   One of the names for the lines connecting nodes in the **link and node** data structure for representing vector data. Others are chain, edge, 1-cell, segment and arc.

Linked list   A data structure for storing lists of items. Each member of the list contains a **pointer** to the next. In a doubly linked list, members also contain pointers to the previous member of the list.

Logarithm   The logarithm of a number n, is the value to which a second number (the base of the logarithm) must be raised to give a value of n. If $n = 2^m$, $m = \log_2(n)$ i.e. m is the logarithm of n with base 2. A logarithmic algorithm is one which is $O(\log n)$ in **big-O notation**.

Logical computational model   A representation of a **conceptual computational model** in a form which is specific to one language or package, but not to any particular make of computer. The data structure used by a particular GIS for vector data may be regarded as a conceptual computational model and the term is synonymous with the use of the term data structure in this book and in the GIS literature.

Mantissa   The actual digits of a **floating point** number.

Minimum Bounding Rectangle   Another name for **Minimum Enclosing Rectangle**.

Minimum Enclosing Rectangle   The smallest rectangle which completely encloses a spatial object.

Monotonic section   Part of a line in which the X or Y coordinates always either increase or decrease.

Morton code   A spatial key for an object formed by alternately taking bits from its X and Y coordinates. When applied to pixels, the key forms the basis of addressing used in **quadtrees**. More generally provides an efficient way of indexing objects by their location.

Node   (1) In GIS, a point joining two lines in the **link and node data structure** for vector data. (2) In graph theory, an item in a **graph** which is connected to other nodes via **edges**. Also referred to as a **vertex**. Tree data structures are often represented using **graphs**, in which case both **leaves** and **branches** are represented by graph nodes.

Planar enforcement   A topological rule which considers all lines in a vector layer to lie in the same plane. This means that lines must intersect at a **node** if they cross. One consequence of this is that if the lines define areas, any point must lie in one of the areas, and can only lie in one area.

Plane sweep algorithm   An algorithm which processes objects in X or Y coordinate order. If objects are processed in decreasing Y order, the action of the algorithm is often visualized as a horizontal line sweeping down the map.

Pointer   Information which connects data in one part of a **data structure** with data elsewhere in the structure.

Polygon   (1) Mathematically, a plane figure with three or more straight sides. (2) In GIS, often used as a synonym for area features, since these are represented as polygons in vector GIS.

Polynomial   A function in which at least one of the terms is raised to a power. For example
$$y = a \cdot x^2 + b \cdot x + c$$

Procedure   In programming, the name for a self-contained section of a program. For example, sqrt(n) might be a procedure, which is passed the value of n as an argument, and returns its square root as a result.

Pseudo-code   A description of an algorithm or heuristic in sufficient detail that a programmer could implement it in any suitable language. This may take the form of a written description in English, but more commonly uses a syntax based on a simplified version of a programming language.

Quadratic   (1) A **polynomial** function in which at least one term is raised to the power of 2, and no terms are raised to a higher power. (2) A quadratic algorithm is one which is O(n^2) in **big-O notation**.

Quadtree   A data structure which stores spatial data by successively halving the area of interest until each element is internally uniform. Commonly applied to raster data, in which the subdivision continues until all pixels within an area have the same value. Can provide compression of raster data and is important as an efficient spatial index.

Recursion   The name given to the style of programming in which a procedure calls itself. Many **algorithms** in which a problem is broken down into successively smaller parts, lend themselves naturally to recursion.

Register   Part of CPU which allows data which is used repeatedly to be stored in the CPU instead of being read from memory each time.

Root   The starting node of a **tree data structure**. Since trees are normally drawn with their **edges** hanging down, the root is usually at the top of the diagram.

Run length encoding   A data structure for the storage of raster data. Instead of storing the value of every pixel, runs of pixels with the same value are identified. The pixel value and length of each run is then stored.

Secondary storage   General term for the storage of data on a permanent medium, such as CD or magnetic disk. Data must be read into memory from secondary storage before it can be used.

Segment   (1) The name used for the lines connecting nodes in the **DIME** data structure for representing vector data. (2) The straight line connecting two points as part of a link.

Single precision   The storage of floating point numbers in a 4 **byte** (32 bit) **word**. This means that calculations only have a **precision** of approximately 6 to 7 significant digits.

Sliver polygon   Small **polygons** formed when two almost identical lines are overlaid. This can arise if the same line is digitized twice by mistake, since the two versions will be similar but will differ slightly in the location of the digitised points.

Toggle   In programming, a variable which flips between two possible values.

Topological data   Data which store information about the connections between vector objects. Sometimes loosely referred to as topology (see **Geometrical data**).

Tree   A data structure consisting of **nodes** or **vertices**, connected by **edges**. The structure has a single **node** which is the **root**. This will normally have two or more child nodes, which in turn may have child nodes of their own. Trees represent data in a hierarchical manner, and the number of levels in the tree is normally $O(\log n)$. Queries which only have to visit each level, rather than each node, will therefore often have $O(\log n)$ **complexity**.

Vertex   (1) In GIS, any of the points along a digitized line except the **nodes** at each end. (2) In graph theory, an alternative term for **node**.

Word   Unit of storage consisting of several bytes. 2 and 4 byte words are commonly used to store integers, and words of 4 or more bytes to store **floating point** numbers.

Bibliography

Abel, D. J. and Mark, D. M. (1990) 'A comparative analysis of some two-dimensional orderings', *Int J. Geographical Information Systems* 4(1): 21–31.

Arge, L. (1997) 'External-memory algorithms with applications in GIS', in M. Van Kreveld, J. Nievergelt, T. Roos and P. Widmayer (eds) *Algorithmic foundations of Geographic Information Systems*, Lecture Notes in Computer Science 1340, Berlin: Springer-Verlag, pp. 213–254.

Berry, J. K. (1993) *Beyond mapping: concepts, algorithms and issues in GIS*, GIS World Books.

Berry, J. K. (1995) *Spatial reasoning for effective GIS*, GIS World Books.

Blakemore, M. (1984) 'Generalization and error in spatial databases', *Cartographica* 21: 131–139.

Bowyer, A. and Woodwark, J. (1983) *A programmer's geometry*, London: Butterworth.

Bugnion, E., Roos, T., Wattenhofer, R. and Widmayer, P. (1997) 'Space Filling Curves and Random Walks', in M. Van Kreveld, J. Nievergelt, T. Roos and P. Widmayer (eds) *Algorithmic foundations of Geographic Information Systems*, Lecture Notes in Computer Science 1340, Berlin: Springer-Verlag, pp. 199–211.

Burrough, P. A. and McDonnell, R. A. (1998) *Principles of Geographical Information Systems*, Oxford University Press.

Cormen, T. H., Leiserson, C. E. and Rivest, R. L. (1990) *Introduction to algorithms*, Cambridge, Massachusetts: MIT Press.

De Berg, M., Van Kreveld, M., Overmars, M. and Schwarzkopf, O. (1997) *Computational geometry*, Berlin: Springer.

De Simone, M. (1986) 'Automatic structuring and feature recognition for large scale digital mapping', *Proc Auto Carto London*, London, AutoCarto London Ltd, pp. 86–95.

Devereux, B. and Mayo, T. (1992) Task Oriented Tools for Catographic Data Capture: *Proceedings AGI'92*, p. 2.14.1–2.14.7, London, Westrade Fairs Ltd.

Douglas, D. H. (1974) 'It makes me so cross', Reprinted in D. J. Peuquet and D. F. Marble (eds) (1990) *Introductory readings in Geographical Information Systems*, London: Taylor and Francis.

Dutton, G. (1999). *A hierarchical coordinate system for geoprocessing and cartography*, Lecture Notes in Earth Science 79, Berlin: Springer-Verlag.

Fisher, P. F. (1993) 'Algorithm and implementation uncertainty in viewshed analysis', *International Journal of Geographical Information Systems* 7(4): 331–347.

Fisher, P. F. (1997) 'The pixel – a snare and a delusion', *Int. J. Remote Sensing* 18(3): 679–685.

Foley, J. D., van Dam, A., Feiner, S. K. and Hughes, J. F. (1990) *Computer graphics: principles and practice*, Reading MA: Addison-Wesley.

Gahegan, M. N. (1989) 'An efficient use of quadtrees in a geographical information system', *International Journal of Geographical Information Systems* 3(3): 201–214.

Gardiner, V. (1982) 'Stream networks and digital cartography', *Cartographica* 19(2): 38–44.

Gold, C. M. (1992) 'Surface interpolation as a Voronoi spatial adjacency problem', *Proceedings, Canadian Conference on GIS*, Ottawa, ON, pp. 419–431 (available on http://www.voronoi.com).

Goodchild, M. F. and Grandfield, A. W. (1983) 'Optimizing raster storage: an examination of four alternatives', *Proceedings Auto Carto 6*, Ottawa, 1: 400–407.

Greene, D. H. and Yao, F. F. (1986) Finite-Resolution Computational Geometry. *Proceedings 27th Symposium on Foundations of Computer Science*, 143–152.

Haggett, P. and Chorley, R. J. (1969) *Network analysis in geography*, London: Edward Arnold.

Healey, R. G. (1991) 'Database Management Systems', in D. J. Maguire, M. F. Goodchild and D. W. Rhind (eds) *Geographical Information Systems – principles and applications*, Harlow: Longman, Vol. 1, pp. 251–267.

Hogg, J., McCormack, J. E., Roberts, S. A., Gahegan, M. N. and Hoyle, B. S. (1993) 'Automated derivation of stream channel networks and related catchment characteristics from digital elevation models', in P. M. Mather (ed.) *Geographical Information handling – Research and Applications*, pp. 207–235.

Huang, C.-W. and Shih, T.-Y. (1997) 'On the complexity of the point in polygon algorithm', *Computers and Geosciences* 23(11): 109–118.

Jenson, S. K. and Domingue, J. O. (1988) 'Extracting topographic structure from digital elevation model data for geographic information system analysis', *Photogrammetric Engineering and Remote Sensing* 54: 1593–1600.

Jones, C. B. (1997) *Geographical Information Systems and Computer Cartography*, Harlow: Longman.

Keating, T., Phillips, W. and Ingran, K. (1987) 'An integrated topologic database design for geographic information systems', *Photogrammetric Engineering and Remote Sensing* 53: 1399–1402.

Kidner, D. B., Ware, J. M., Sparkes, A. J. and Jones, C. B. (2000) 'Multiscale terrain and topographic mapping with the implicit TIN', *Transactions in GIS* 4(4): 379–408.

Knuth, D. (1998) *Sorting and Searching* (Volume 3 of The Art of Computer Programming), 2nd edn, Reading MA: Addison-Wesley.

Kumler, M. P. (1994) 'An intensive comparison of Triangulated Irregular Networks (TINs) and Digital Elevation Models (DEMs)', *Cartographica* 31(2).

Lam, N. (1983) 'Spatial Interpolation Methods: A Review', *The American Cartographer* 10(2): 129–149.

Longley, P. A., Goodchild, M. F., Maguire, D. J. and Rhind, D. W. (eds) *Geographical Information Systems: Principles, Techniques, Applications and Management*. London, John Wiley and Son.

Maguire, D. J., Goodchild, M. F. and Rhind, D. W. (1991) *Geographical Information Systems – Principles and Applications (2 Volumes)*, Harlow: Longman. Many chapters now online at http://www.wiley.com/gis under 'Links to Big Book 1'.

Mark, D. M. (1984) 'Automated detection of drainage networks from digital elevation models', *Cartographica* 21: 168–178.

Mark, D. M. (1975) 'Computer analysis of topography: a comparison of terrain storage methods', *Geografisker Annaler* 57A: 179–188.

Mark, D. M. (1979) 'Phenomenon-based data-structuring and digital terrain modelling', *Geo-Processing* 1: 27–36.

Marks, D., Dozier, J. and Frew, J. (1984) 'Automated basin delineation from digital elevation data', *Geo-Processing* 2: 299–311.

Mather, P. M. (1999) *Computer Processing of Remotely-Sensed Images: An Introduction*, 2nd edn, Chichester: Wiley.

Nagy, G. (1980) 'What is a 'good' data structure for 2-D points'? in H. Freeman and G. G. Pieroni (eds) *Map Data Processing*, New York: Academic Press.

Nievergelt, J. (1997) 'Introduction to Geometric Computing: from Algorithms to Sofware', in M. Van Kreveld, J. Nievergelt, T. Roos and P. Widmayer (eds) *Algorithmic foundations of Geographic Information Systems*, Lecture Notes in Computer Science 1340, Berlin: Springer-Verlag, pp. 1–19.

Nievergelt, J. and Widmayer, P. (1997) 'Spatial Data Structures: Concepts and Design Choices', in M. Van Kreveld, J. Nievergelt, T. Roos and P. Widmayer (eds) *Algorithmic foundations of Geographic Information Systems*, Lecture Notes in Computer Science 1340, Berlin: Springer-Verlag, pp. 153–197.

Peucker, T. K., Fowler, R. J., Little, J. J. and Mark, D. M. (1978) 'The Triangulated Irregular Network', *Proceedings American Society of Photogrammetry Digital Terrain Models Symposium*, St. Louis, MO, pp. 516–540.

Peucker, T. K. and Chrisman, N. (1975) 'Cartographic Data Structures', *American Cartographer* 2(1): 55–69.

Peuquet, D. J. (1981a) 'An examination of techniques for reformatting digital cartographic data, Part I, The raster-to-vector process', *Cartographica* 18(1): 34–48.

Peuquet, D. J. (1981b) 'An examination of techniques for reformatting digital cartographic data, Part II, The vector-to-raster process', *Cartographica* 18(3): 21–33.

Peuquet, D. J. and Marble, D. F. (1990) *Introductory Readings in Geographical Information Systems*, London: Taylor and Francis.

Peuquet, D. J. (1984) 'A conceptual framework and comparison of spatial data models', *Cartographica* 21(4): 66–113.

Rosenfeld, A. (1980) 'Tree structures for region representation', in H. Freeman and G. G. Pieroni (eds) *Map Data Processing*, New York: Academic Press.

Rosenfeld, A. and Kak, A. (1982) *Digital picture processing*, London: Academic Press.

Saalfeld, A. (1987) 'It doesn't make me nearly as CROSS', *International Journal of Geographical Information Systems* 1(4): 379–386.

Samet, H. (1990a) *Applications of spatial data structures: computer graphics, image processing, and GIS*, Reading, MA: Addison-Wesley.

Samet, H. (1990b) *The design and analysis of spatial data structures*, Reading, MA: Addison-Wesley.

Samet, H. and Aref, W. G. (1995) 'Spatial data models and query processing', in W. Kim (ed.) *Modern database systems: the object model, interoperability and beyond*, Addison-Wesley/ACM Press, pp. 338–360.

Schirra, S. (1997) 'Precision and Robustness in Geometric Computations', in M. Van Kreveld, J. Nievergelt, T. Roos and P. Widmayer (eds) *Algorithmic foundations of Geographic Information Systems*, Lecture Notes in Computer Science 1340, Berlin: Springer-Verlag, pp. 255–287.

Shaffer, C. A., Samet, H. and Nelson, R. C. (1990) 'QUILT: a geographic information system based on quadtrees', *International Journal of Geographical Information Systems* 4(2): 103–131.

Sibson, R. (1981) 'A brief description of natural neighbour interpolation', in V. Barnett (ed.) *Interpreting Multivariate Data*, Chichester: Wiley, pp. 21–36.

Skidmore, A. K. (1989) 'A comparison of techniques for calculating gradient and aspect from a gridded digital elevation model', *International Journal of Geographical Information Systems* 3(4): 323–334.

Smith, T. R., Menon, S., Star, J. L. and Estes, J. E. (1987) 'Requirements and principles for the implementation and construction of a large-scale geographic information system', *International Journal of Geographical Information Systems* 1(1): 13–31.

Tarboton, D. G. (1997) 'A new method for the determination of flow directions and upslope areas in grid digital elevation models', *Water Resources Research* 33(2): 309–319.

Teng, T. A. (1986) 'Polygon overlay processing: a comparison of pure geometric manipulation and topological overlay processing', *Proceedings 2nd International Symposium on Spatial Data Handling*, pp. 102–119.

Theobald, D. M. and Goodchild, M. F. (1990) 'Artifacts of TIN-based surface flow modelling', *Proceedings GIS/LIS '90*, vol. 2, ASPRS/ACSM, Bethesda, Maryland, pp. 955–967.

Van Kreveld, M. (1997) 'Digital elevation models and TIN algorithms', in M. Van Kreveld, J. Nievergelt, T. Roos and P. Widmayer (eds) *Algorithmic foundations of Geographic Information Systems*, Lecture Notes in Computer Science 1340, Berlin: Springer-Verlag, pp. 37–78.

Van Kreveld, M., Nievergelt, J., Roos, T. and Widmayer, P. (1997) (eds) *Algorithmic foundations of Geographic Information Systems*, Lecture Notes in Computer Science 1340, Berlin: Springer-Verlag.

Van Oosterom, P. (1999) 'Spatial access methods', in P. A. Longley, M. F. Goodchild, D. J. Maguire and D. W. Rhind (eds) *Geographical Information Systems: principles, techniques, applications and management*, London: John Wilsey and Son, Vol. 1, pp. 385–400.

Waugh, T. C. (1986) 'A response to recent papers and articles on the use of quadtrees for geographic information systems', *Proceedings 2nd International Symposium on Spatial Data Handling*, pp. 33–37.

White, D. (1978) 'A design for polygon overlay', in G. Dutton (ed.) *First International Advanced Study Symposium on Topological Data Structures*, vol. 6.

White, M. (1984) 'Tribulations of automated cartography and how mathematics helps', *Cartographica* 21: 148–159.

Wilson, J. P. and Gallant, J. C. (2000) (eds) *Terrain Analysis – principles and applications*, Chichester: John Wiley and Sons.

Wise, S. M. (1988) 'Using Contents Addressable Filestore for rapid access to a large cartographic database', *International Journal of Geographical Information Systems* 2(2): 11–120.

Wise, S. M. (1995) 'Scanning thematic maps for input to geographic information systems', *Computers and Geosciences* 21(1): 7–29.

Wise, S. M. (1998) 'The effect of GIS interpolation errors on the use of DEMs in geomorphology', in S. N. Lane, K. S. Richards and J. H. Chandlen (eds) *Landform Monitoring, Modelling and Analysis*, Chichester: Wiley, 139–164.

Wise, S. M. (1999) 'Extracting raster GIS data from scanned thematic maps', *Transactions in GIS*, pp. 221–237.

Wise, S. M. (2000) 'Data modelling in GIS – lessons from the analysis of Digital Elevation Models', (Guest editorial) *International Journal of Geographical Information Science* 14(4): 313–318.

Worboys, M. F. (1995) *GIS: A computing perspective*, London: Taylor and Francis.

Yu, S., Van Kreveld, M. and Snoeyink, J. (1996) 'Drainage queries in TINs: from local to global and back again', in M.-J. Kraak and M. Molenaar (eds) *Advances in GIS II: Proceedings 7th Int. Symposium on Spatial Data Handling*, London: Taylor and Francis, pp. 829–855.

Index